Law and Society

Recent Scholarship

Edited by Eric Rise

A Series from LFB Scholarly

Sorting Out Deregulation

Protecting Free Speech and Internet Access in The United States, Germany, and Japan

Jae-Young Kim

LFB Scholarly Publishing LLC
New York 2002

Library of Congress Cataloging-in-Publication Data

Kim, Jae Young, 1968-
Sorting out deregulation : protecting free speech and Internet access in the United States, Germany, and Japan / Jae-Young Kim.
p. cm. -- (Law and society)
Includes bibliographical references and index.
ISBN 1-931202-37-0 (alk. paper)
1. Freedom of speech. 2. Internet--Access control. 3. Telecommunication--Deregulation. I. Title. II. Law and society (New York, N.Y.)
K3254 .K56 2002
323.44'3--dc21

2002007182

ISBN 1-931202-37-0

Printed on acid-free 250-year-life paper.

Manufactured in the United States of America.

Table of Contents

CHAPTER 1

Introduction

It is a truism that the movement toward telecommunications deregulation is a worldwide phenomenon, even though there is little or no general pattern. This study examines its definitive or probable effects on two significant public goods, universal service and free speech on the Internet, in the United States, Germany, and Japan. Because the experiences of making telecommunications policy in these three countries differ in many aspects, and each has some unique features, they likely provide useful lessons for the rest of the world, especially for countries that are in the initial stage of developing or restructuring their telecommunications infrastructures. The study examines whether deregulation can be reconciled with public interest values and explores the changing role of the state in developing a new communication environment.

Defining the Problem and Justifications

For almost a century, the term, "telecommunications," was synonymous with telephone systems. Today, it covers all forms of messages – digital data, video, and audio – transmitted through a variety of electronic means as well as the old wires. Since the introduction of the telegraph as the first telecommunications medium, the development and utilization of telecommunications have been key factors in the progress of modern civilization. Nevertheless, as Garnham (1992) pointed out, "the study of the media has excluded the study of telecommunications, leaving the field to engineers and economists" (p. 339).[1] This is no longer the case today, however, as telecommunications is increasingly forcing its way onto the mass communication studies agenda.[2]

The most significant reason for such change is the coming together of all forms of mediated communications into an electronic, digital form, processed by computers. This convergence is integration between or among different media systems – the merger of telephone, data transmission, and computer technologies resulting into a fully integrated telecommunications system. It is a process wherein the storage and retrieval of text, data, sound, and image occur within a single, integrated environment (Dennis & Pavlik, 1993, p. 2). The development of digital communication, represented by a simple on and off, - and +, or 0 and 1 signal, is the fundamental basis for this convergence.

The technological convergence of media, telecommunications, and computer industries makes "the resultant sector the largest and fastest growing component of the global economy" (McChesney, 1996a, p. 5). Indeed, telecommunications plays a fundamental role in the future well-being of every country's economy. It is not only the direct contribution of the sector to the economic growth but also its crucial role in supporting economic activity that make telecommunications so important. But convergence also brings about uncertainty about the future for the industry. In this respect, as Mosco (1984) mentioned much earlier, "to do policy research in telecommunications today is to address the major issues of our time" (p. ix). Telecommunications policy research is both relevant and urgent in the field of mass communications.

This study examines recent changes in telecommunications policy. The historical rationale behind telecommunications regulation is that the function of telecommunications is imbued with broader public interest or equity-based values. According to Horwitz (1989), telecommunications, together with transportation, energy utilities, and the currency exchange system, is the essential mode that permits trade and discourse among members of a society and provides the services upon which all economic activity depends (p. 11). Therefore, the construction and maintenance of telecommunications infrastructure have usually been the responsibility of governments. Governments have had to guarantee that the provision of telecommunications services corresponded to the principle of equal opportunity for the suppliers and recipients alike, that any type of concentration and manipulation was prevented, and that social integration was not endangered. In this sense, telecommunications policy was a public task.

The current environment, based on faith in market-based solutions to economic problems, is referred to variously as "privatization," "deregulation," "commercialization," and "liberalization."[3] These terms also sum up the dominant economic trend worldwide. In the United States, deregulation of various key industries began with the deregulation of airlines by the Carter Administration in the late 1970s. The trucking, natural gas, railroad, and telecommunications industries followed in the 1980s and 1990s. Financial services and electric power are among industries currently being deregulated.[4]

Deregulatory movements initiated in the late 1970s have spawned an age in which market criteria gain ground in all areas, and competition becomes an independent regulator of the development of telecommunications. The 1980s had seen a shift toward a market model of telecommunications in the United States and Western Europe. Additionally, in the post-Communist regimes of Eastern European and in most less developed countries, "market liberalism enjoys a stunning rise of popularity" (Keane, 1991, pp. x-xi). Whether one regards it as a good or bad development, it is evident that this is happening.

However, because of its wide variety in stages or paths, there is no general or universal explanation for the current global trend toward telecommunications deregulation. For example, while many of the less developed nations are now selling off industries that had previously been government-owned, the more advanced countries of Western Europe and North America are loosening up already liberalized economic regimes. Indeed, as Hancher & Moran (1989) put it, "the most striking feature of deregulation is variety – in incidence, form and extent" (p. 129). To investigate the deregulatory trend as a phenomenon, one must identify the reasons why a specific society is inclined to deregulate its telecommunications. Broad-based generalizations or conclusions can be drawn only after beginning with these case studies.

The telecommunications deregulatory movements in the United States, Germany, and Japan are particularly worthy of attention, not only because they have advanced telecommunications systems, but because they have each taken different paths in developing their telecommunications sectors. The United States has the most sophisticated telecommunications system, and its regulatory framework rests on private ownership and market incentives. By establishing the first and the most conspicuous telecommunications deregulation, the United States could serve as a model for other less developed countries, though its experience has by no means been uniformly positive or affirmative.

On the other hand, the development of telecommunications policy in Germany has been marked by a relatively skeptical attitude toward "the workability of the market-economy system and ... the possibility of catering to public benefits via the market mechanism" (Hoffman-Riem, 1984, p. 55). Perhaps because the weight of German history has favored retaining the public monopoly model along with its unique constitutional restrictions, Germany lags behind in telecommunications reform among European Union (EU) countries who are undergoing telecommunications liberalization efforts and recognizing the importance of competition in this area. As Europe' biggest telecommunications market, Germany is considered a microcosm of the broader EU market and the principal test case for the liberalization of telecommunications services.

Japan liberalized its telecommunications after the United States but leads Asia in deregulatory efforts. Since 1985 when the telecommunications industry was substantially reformed, Japan has shifted from a government monopoly to market competition. However, this trend proceeds in Japan in a manner different from that in the United States and Germany because of various cultural, legal, and social differences between East and West. Japan has much in common with other Far Eastern Asian countries regarding political, economic, social, and cultural respects including a traditionally slower pace of administrative reform. The Confucian commitment also maintains cohesion of the social system. With due consideration of this distinctiveness, Japanese telecommunications policy is expected to be a useful laboratory for the policy-makers who are considering a non-Western type of telecommunications development.

In short, the three countries' experiences in telecommunications policy are selected for thorough examinations for these reasons: (1) early adoption of telecommunications deregulation with a tradition of private ownership in the United States, (2) adoption within regional alliance and with a history of public ownership in Germany, and (3) adoption within Eastern cultural context in Japan.

The main reason that this study deals with the Internet instead of one of the many other new media is due to the discourse concerning "electronic democracy."[5] Electronic democracy is a vision of a new electronic environment where direct, participatory democracy is possible in previously unimaginable ways. Although regular assemblies of the people as a whole are technically impossible in large-scale, complex societies, the Internet enables modern democracy grounded in both mechanisms of representation and the institutional division of state and civil society to advance into a direct form of governance like *agora*.[6] Likewise, the concept of electronic democracy "connotes a radically new form of democratic practice modified by new information technologies" (Friedland, 1996, p. 185).

Expectations for electronic democracy rely heavily on the uniquely open structure and affordable cost of the Internet (Berman & Weitzner, 1997). Even if traditional media have been readily available around the world, they have failed to enable full democratic participation because of architectural limitations. In contrast, the Internet's interactive and decentralizing quality allows for untrammeled expression, a diversity of views, and exchange of opinions in a nonhierarchical manner that are simply impossible in any other communications medium. In this vein, Baran (1996) described the burgeoning Internet as "an example of a potentially positive development emerging from the so-called telecommunications revolution," (p. 60) which could pave the way for a future democratic society.

However, the Internet is a new information technology with a double-edge. It can be designed and operated in ways that either push people further apart or bring them closer together. The outcome largely depends on the policies that distribute the benefits of new technology in a society. Without a good policy guiding them, the opportunity for electronic democracy is likely to wind up as a kind of "vaster wasteland."[7]

To realize the promise of electronic democracy there are certain objectives that must be satisfied. Above all, universal service and free speech are significant public goods worthy of full guarantee as long as the formation of democracy is a cardinal goal of telecommunications policy. As for universal service, it is essential to ensure that all members of society have not only the greatest possible but also equal access to channels of information. In particular, when the benefits of using the Internet flow to higher income and education attainment groups at an accelerating rate, a universal service policy that encourages a leveling between information haves and have-nots needs to be developed if electronic democracy is to evolve (Civille, 1995, pp. 194-195).

At the same time, it is crucial to enhance an "uninhibited, robust, and wide-open" (*New York Times, Co. v. Sullivan*, 1964, p. 270) discourse in a new electronic environment. The importance of free speech for democracy was well demonstrated in the United States Supreme Court in *Marsh v. Alabama* (1946) in which Justice Black noted that "to act as good citizens they must be informed. In order to enable them to be properly informed their information must be uncensored" (p. 508).

With the advance of deregulation in telecommunications, however, the relationship between the market and public interest has been the subject of many disputes. The proponents of the marketplace model do not downplay the significance of public interest values in telecommunications. Instead, they state that public interests are best served through market forces and competition. This market-oriented perspective of the public interest is at the heart of contemporary telecommunications deregulation.

Specifically, universal service was neither complicated nor controversial under the monopoly telephone systems in Europe and Japan, where they were a public monopoly, or in the United States where a private monopoly was established. Technological advances in communication media and the introduction of competition, however, change monopolistic market structures, and the traditional support for universal service is undermined (Hadden & Lenert, 1995, p. 121; Sugaya, 1997, p. 179). The consequences of telecommunications deregulation also are closely related to the amount of protection for electronic free speech because telecommunications "constitutes the public realm of ideas and discussion, and hence implicates the range of issues surrounding freedom of speech" (Horwitz, 1989, p. 8). The relationships between telecommunications deregulation and two public interests on the Internet, universal service and free speech, are direct and critically important.

Even though one of the most important issues in telecommunications policy is where and how the two conflicting requirements – introducing deregulation and attaining public interests – should be reconciled, not enough systematic thought has been given to the topic. The primary concern of this study is to examine definitive or potential effects on public interests concerning the Internet that might result after a restructuring of the telecommunications industry.

Theoretical Concerns

This study uses political economy[8] as the theoretical framework to inquire about the impact of telecommunications deregulation on public interest values on the Internet. Political economy is a perspective that searches for answers to the question of the power of communication in the analysis of control structures. Communication is defined as "a social process of exchange whose product is the mark or embodiment of a social relationship," (Mosco, 1996, p. 72) and control refers to "expressions of dynamic processes and power relations in the entire social system" (Mosco, 1988, p. 117).

In the field of mass communications, political economy has developed as an alternative way of thinking to the mainstream's theoretical and methodological concerns with "effects," especially in the United States. Communication researchers have recently assembled an impressive array of political economic analyses of media industries and telecommunications policies.

According to Mosco (1996), political economy is defined, in part, by setting itself apart from various theories of telecommunications regulation and the ensuing policy perspectives (p. 81). Three distinguishing features of political economy are identified in the process. First, it is holistic in that a particular aspect of social phenomena is guided by a wider social totality. Second, social or class power is essential in constituting the social whole. Third, political economy is concerned with normative or moral values. These theoretical principles enable this study to explore how power relations in government and in marketplace change with the deregulation of telecommunications and how the relations affect the public interest portion of Internet policy. They also provide a structure for organizing the primary sources used for this study in a systematic way.

Research Questions

The main objective of this study is to examine and compare efforts made by the United States, Germany, and Japan to ensure universal service and free speech on the Internet in the course of establishing deregulatory telecommunications policies. Because each country has a sophisticated, yet uniquely developed telecommunications system, the deregulatory process in each can provide useful suggestions for the role of the state in transforming electronic democracy.

Before dealing with that issue, this study generates a broad social, political, and economic history of telecommunications policies with a focus on the dynamics of deregulatory movements in three developed countries. Also of great consideration is the identification of universal service and free speech on the Internet as essential public interest values in the construction of electronic democracy. After addressing those matters, the study analyzes the impact of telecommunications deregulation on the two public goods concerning the Internet and discusses findings in relation to the changing role for the state. There are four elemental research questions:

1. What are the key factors affecting telecommunications deregulation in the United States, Germany, and Japan?
2. How has telecommunications deregulation affected universal service on the Internet in the United States, Germany, and Japan?
3. How has telecommunications deregulation affected Internet users' free speech rights in the United States, Germany, and Japan?
4. How should the role of the state be positioned in constructing electronic democracy?

Chapter Outline

This book consists of eight chapters. Chapter two provides a historical context of telecommunications policies in the United States, Germany, and Japan. Their privatization efforts are described in detail. Next the theoretical framework for this study is discussed in relation to various theories of regulation and policy perspective. It concludes by presenting what political economy research on telecommunications addresses and how it has evolved.

Chapter three describes the scope, method, and sources of this study. Although secondary sources are used for the study, the chapter mainly presents the availability and significance of primary sources.

Chapter four specifically identifies the causes of the telecommunications deregulation trends in the United States, Germany, and Japan. The motives are analyzed in both common and individual respects, and each case is compared to the others.

Next come two main chapters containing the analyses of consequences of telecommunications deregulation on universal service and free speech rights on the Internet in the United States, Germany, and Japan. Each chapter identifies the imperative values on the Internet for electronic democracy in liberalized market conditions by explicating the history of the Internet and its technological characteristics.

Chapter seven synthesizes all discussions in the previous two chapters and builds on the conclusion of the impact of telecommunications deregulation on public interest. By placing deregulation into the relationship between social control and market from the perspective of political economy, the chapter also reconsiders the role of the state in a current social environment where telecommunications, including the Internet, is increasingly commercialized, privatized, and deregulated.

The last chapter briefly summarizes the major findings of this study. It also suggests implications of this study for other countries.

CHAPTER 2

Review of Literature

This chapter is composed of three sections. The first section presents a historical overview of telecommunications policies as they have evolved in the United States, Germany, and Japan with a focus on their deregulatory efforts. The second section builds a theoretical framework for this study. To examine the impacts of telecommunications deregulation on public interest values, this study takes political economy as a primary analytical framework. The last section reviews scholarly literature on telecommunications from the perspective of political economy and highlights the chief issues and findings that are debated and produced.

Evolution of Telecommunications Policies

Telecommunications policy has changed from regulation to liberalization in many countries since the early 1980s. The reasons and processes of the change differ from nation to nation depending on the country's politic-economic environment. This section examines both how telecommunications policies have evolved within the three countries' econo-political circumstances and how the resulting policies have affected the development of each country's telecommunications system.

Spearheading a Private Model: The American Case

From its inception, United States telecommunications has evolved differently from that of most other countries. Whereas European governments operated telecommunications directly, mainly in association with the post office, the United States government allowed private companies to operate as a regulated monopoly. Since the beginning, the United States has relied on competitive forces to spur technological innovation and to enhance telecommunications services. But during the last two decades, the United States telecommunications industry has evolved from one dominated by a handful of closely regulated firms[1] to one in which a wide variety of providers using a multitude of new technologies compete to meet the increasingly sophisticated communications needs of businesses and consumers.

Bell Telephone received the first patent for a telephone in 1876 and staved off Western Union's bid to transform itself into a phone company. Through a series of acquisitions and mergers, Bell became the major avenue of growth for American Telephone and Telegraph (AT&T), the largest private monopoly providing telecommunications services. Since the early days, the bulk of telecommunications services had been supplied by AT&T and its member operating companies. As a holding company, AT&T controlled stock in 24 operating telephone companies located throughout the United States.

However, the industry was not completely dominated by the Bell System (companies owned by or affiliated with AT&T), but divided between it and the independents (nonaffiliated with AT&T, but often components of large non-Bell holding companies). The expiration of the original phone patent in 1893 allowed many independent phone companies to enter the market, providing phone service and selling equipment. Although the independents covered about 85 percent of the geographical area of the United States, the Bell System served 80 percent of the telephones.

This early rivalry between the Bell and independent interests led to a headlong race to wire the country. However, telephone competition posed a connectivity problem in the early 1900s because the competing exchanges refused to interconnect among telephone users. In 1913, concerned over Bell's acquisition of independents, the Department of Justice (DOJ) filed its first antitrust suit against AT&T. The Kingsbury

Commitment was made by AT&T Vice President Nathan Kingsbury's letter to the Attorney General. In it, AT&T pledged to dispose of its stock in Western Union, to cease acquisition of independents, and to allow independents to interconnect into the Bell System ("Kingsbury Commitment," 1913). In exchange, the DOJ relaxed its pressure on AT&T.

Additionally, after a debate on the advantages of competition and monopoly, Congress determined the consumer would be best served by a single monopoly system that provided universal service, as long as it was required to interconnect all telephone subscribers. The Willis-Graham Act[2] in 1921 forced competing local exchanges to merge into territorial monopolies and to link into a nationwide system. In turn, the Act exempted telephone companies from the antitrust laws (Mueller, 1997a, pp. 40-41; Mueller, 1997b, p. 657).

Congress' favor of this single monopoly telephone system was continuously reflected in the Communications Act of 1934 (1934 Act).[3] By establishing the Federal Communications Commission (FCC), the 1934 Act granted AT&T immunity from antitrust actions in return for a promise to serve the public interest by providing reliable services at affordable prices. However, after enjoying government protection during the first half of the century, the AT&T's monopoly structure gradually "proved an impediment to growth and innovation" and "fell in disfavor with regulators for [its] inefficient and anti-competitive market conduct" (Atkin, 1996, p. 233). This ultimately resulted in the breakup of AT&T, an early turning point in the history of United States telecommunications policy.

Since the 1950s, the monopolies had begun to break, with a series of regulatory decisions, court cases, and technological innovation. In 1949, the DOJ, based on complaints from state public utilities commissions, filed suit against Western Electric and AT&T for violations of the Sherman Act in the manufacture and sale of telephone equipment and supplies. This antitrust suit was settled in a 1956 Consent Decree, which constructed the appropriate boundary between regulated monopoly communications activities and unregulated competitive data processing services. The Consent Decree declared that AT&T could engage only in the business of providing regulated common carrier telecommunications services. This agreement also stated that the DOJ would drop its lawsuit if AT&T agreed to stay out of the fledgling computing industry.

The first technological challenge to the monopolies was concerned with Hush-A-Phone, a device that fits over the telephone receiver for user privacy. On the pretext of incompatibility and unreliability, AT&T did not permit any "foreign attachment" to its network including that device. Hush-A-Phone Corporation petitioned the FCC to vacate this prohibition. On appeal, the court held that AT&T's restriction was "an unwarranted interference with the telephone subscriber's right reasonably to use his telephone in ways that are privately beneficial without being publicly detrimental" (*Hush-A-Phone Corp. v. United States*, 1956, p. 269).

A similar decision was made in 1968, when the FCC ruled against AT&T's ban of an acoustic coupler connecting radiotelephones called Carterfone. The Carterfone device allowed mobile telephone users to access the public telephone network by connecting the national telephone landline system with two-way mobile radios. Relying on the principles of *Hush-A-Phone* (1956), and finding that there was no potentially harmful interference to the network and that the device satisfied an unmet demand, the FCC ruled AT&T's prohibition of Carterfone unlawful (FCC, 1968). This case, along with the *Hush-A-Phone* precedent, contributed to opening the telephone attachment market, in spite of AT&T's effort to restrict competition. Moreover, it provided small firms with an incentive to innovate new telephone equipment of many kinds.

During the following year, the evolution of transmission technologies brought challenges to AT&T's monopoly, resulting in the permission for Microwave Communications, Inc. (MCI) to offer private long distance service over its own network. This case was originally initiated in 1963, when MCI, a fledgling company, tried to construct a microwave line between Chicago and St. Louis to meet the communications needs of small businesses. At that time, however, the private systems could be built by very large users only (FCC, 1959). In a 4-3 vote, the FCC approved the MCI application, mainly because MCI was expected to capture a latent market (FCC, 1969). It became the first step to competition in long-distance markets.

Soon the advance of communication satellite and computer communications challenged the natural monopoly theory. After long and heated debate over the potential and management of satellites in the 1960s, the United States adopted the so-called "Open Skies" policy in 1972 (FCC, 1972). Almost all private companies were free to establish satellite communications networks. AT&T was only permitted to lease satellite circuits, because its market strength might allow it to engage in cross-subsidization,[4] thereby forestalling technological innovation. The policy established satellite communications as an open, competitive industry rather than as part of the AT&T monopoly.

As the distinction between computer and communications technologies became less clear, the FCC attempted to deal with differences between basic and so-called enhanced services.[5] To prevent harmful discrimination by the carriers (such as the prevention of other enhanced services providers from accessing their own networks) the FCC required carriers to separate their enhanced from their basic services business (FCC, 1986).

The direct effect on the divestiture of AT&T was the application of the antitrust challenge in 1982. AT&T was already accused of abusing its monopoly power several times, largely by refusing to interconnect with other carriers or by pricing below cost to drive competitors out of the market. As described above, AT&T was the subject of major antitrust laws in 1913 and 1956. Nonetheless, AT&T could retain its monopoly status in return for subjecting itself to regulation under the Kingsbury Commitment of 1913 and under the Consent Decree of 1956.

This arrangement lasted until January 8, 1982, when the DOJ and AT&T jointly announced that they had reached a settlement to divest AT&T's local operating companies. The settlement originated in a lawsuit filed by MCI in 1974 when the pressure for competition in the long-distance and telephone equipment industries was just beginning to heat up. MCI, challenging the underlying economic rationale behind the natural monopoly doctrine, asserted that the Bell System had illegally restrained market competition. Despite AT&T's attempts to delay a final decision by the DOJ and circuit courts, it had become evident that the monopoly theory would not prevent an eventual breakup of its system. The lawsuit led to a 1982 Consent Decree that would become the single most important movement toward telecommunications deregulation.

In a district court, a Modified Final Judgment (MFJ) was adopted and implemented in January 1984 (*United States v. AT&T*, 1982). Under the MFJ, AT&T's 22 local telephone companies were divested into seven independent Regional Bell Operating Companies (RBOCs).[6] These "Baby Bells" were not allowed into the competitive markets of long distance telephone service, equipment manufacture, or data processing. On the other hand, AT&T was no longer in the business of providing basic local switched telephone service, although it was allowed to retain several major subsidiaries[7] and enter all domestic and international competitive markets. In particular, AT&T was permitted to enter the computer and ancillary industries since the MFJ removed restrictions imposed by the 1956 Consent Decree.

The divestiture of AT&T undermined many of the traditional principles of United States telecommunications policy. It was closely linked to a broader trend in that the breakup was "used to expand the program of the deregulation of telecommunications" (Horwitz, 1989, p. 244). The decade following divestiture saw competitive market forces dramatically reshape the telecommunications landscape.[8] Especially in areas where competition was introduced by the MFJ, new services were brought to market by competitors and consumer prices largely dropped. Shaw (1998) defined the period from the mid-1980s through 1995 as "an era in which, through initial experimentation with deregulation, the nation would experience unprecedented services" (p. 35).

The deregulatory trend reached a peak with the passage of the Telecommunications Act of 1996 (1996 Act).[9] The foremost purpose of the Act was to deregulate all telecommunications industries and to permit the market to determine the telecommunications system. Until 1996, federal regulations kept telephone, cable television, and broadcast television industries separate from each other. The 1996 Act, however, changed the playing field considerably. It allowed cross-ownership across these media in an effort to encourage competition and technological innovation. Indeed, the local telephone markets, which had been serviced by the seven independent, but monopolistic RBOCs, were completely opened to the long-distance carriers and cable operators. Likewise, by opening local telephone markets to competition, the 1996 Act supersedes the MFJ on the AT&T divestiture.

With the lifting of the cross-ownership restriction, traditional cable companies also can start local phone businesses without any franchise. Both phone and cable companies may opt to create "open video systems" – pay television services – using any communications technology, by opening two-thirds of the channels on their systems to outsiders on an equal basis. In a word, the 1996 Act marks the biggest overhaul of United States telecommunications policy since the 1934 Act. The boundaries recognized and codified between broadcasting and telephony in the 1934 Act were broken down.[10]

A Laggard Moves Toward Deregulation: The German Case

When the telegraph and the telephone were first introduced throughout Europe in the latter part of the 19th century, there was a strong belief that telecommunications was a natural monopoly and that economies of scale were such that competition did not make sense. Therefore, telecommunications in most European countries was dominated by government-owned monopolies. Administration of telegraph and telephone networks was generally merged with the existing postal authority, which had had monopoly over mail for generations. The new administration became known as the PTTs, for Post, Telegraph, and Telephone.[11]

While the United States telecommunications industry grew from a patchwork of independent operators dominated by the early Bell System, into a *de facto* monopoly as the Bell System squeezed out or acquired its competitors, and finally into a regulated monopoly, the state's PTTs were granted a complete monopoly for the provision of telecommunications infrastructure and services. In Germany, the *Deutsche Bundespost* (DBP) was responsible for all telephone services and almost the entire telecommunications network, as well as for the postal service. European PTTs generally became large and powerful organizations – often the largest employers in their respective countries[12] – even though the specific structure of the sector has varied from one country to another.

Many deregulatory reforms have been implemented in the hope of stimulating innovation and efficiency in this sector. Presently, the PTTs of Europe are experiencing unprecedented change. A major influence has been the divestiture of AT&T in the United States, which has called into question the prevalent public concept of natural monopoly. Additionally, the privatization and liberalization of telecommunications in the United Kingdom beginning in 1984 provided a clear alternative to the monopoly model.

More broadly speaking, European countries' approaches to telecommunications restructuring have been boosted by the European integration process, aimed at creating a single common market.[13] Since the European Union (EU) became fully aware of the enormous strategic importance of telecommunications in a global economy and recognized that the inefficiencies and high prices plaguing European telecommunications are detrimental to industrial development, it has developed itself into a major force in the reform of European telecommunications. Every telecommunications policy in Europe has to be balanced with new European policies. The remainder of this subsection reviews the major initiatives toward telecommunications deregulation launched by the EU and then describes German telecommunications from within that context.

Providing Guidelines: The European Union

The major policy-making institutions at the EU level include the Council of Ministers, the European Commission, and the European Parliament.[14] The Commission was empowered by Article 90 of the 1957 Treaty of Rome to issue "directives"[15] that introduce or safeguard competition in sectors controlled by public enterprises that have been granted exclusive rights by national governments. Indeed, the Council plays a major role in guiding the various treaties signed by member states of the EU, in initiating policies, and in proposing legislation to the European Council. It also has the authority to control and regulate anti-competitive practices. In particular, the role of the European Commission in the telecommunications sector can be classified into three main areas: the formulation of the political goals of European regulation, the organization of research and development in the field of information and communications technologies, and the establishment of technical standards (Esser & Noppe, 1996).

Since the mid-1980s when the European Commission listed the telecommunications sector as an industrial policy priority, telecommunications reforms in Europe have been strongly guided by a general blueprint in the form of directives, decisions, resolutions, recommendations, and community legislation. The most important Commission policy document regarding telecommunications is the *Green Paper* (Commission, 1987). The principles laid out in it can be summarized as liberalization and harmonization. By illuminating the major characteristics of the future market structure in European telecommunications, the paper identified areas in which member states would have to take action to achieve the common goals, and it proposed how liberalization and privatization of member state telecommunications monopolies could take place. Specifically, four major goals were set up in the *Green Paper*: (1) opening of terminal equipment markets to competition, (2) progressive introduction of competition in services, (3) separation of regulatory and operational activities, and (4) movement toward more cost-based tariffs.

In accordance with the 1987 *Green Paper*, the market for terminal equipment was liberalized in 1988 when the Commission laid down the framework for the introduction of Europe-wide competition in the provision of terminal equipment in a directive (Commission, 1988). This Terminal Equipment Directive opened the market by abolishing all special or exclusive rights to sell terminals. Moreover, it required all member states to publish their specifications and approval rules for terminals to facilitate foreign supplier entry into the market.

The next significant steps came with two directives approved on June 28, 1990. One is the Services Directive, a cornerstone for liberalizing the European telecommunications market (Commission, 1990). The other is the Open Network Provision (ONP) Directive, which established the framework for access to and use of telecommunications networks and services (Council, 1990).

The Services Directive initiated an opening up to competition of the telecommunications services by providing for the removal of special or exclusive rights granted by member states for value-added services and data services. However, key basic services such as public voice telephony were temporarily permitted to remain under monopoly control. It also established deadlines for full liberalization and required the separation of operational and regulatory functions.

The 1990 ONP Directive sought to ensure open access to and harmonization of conditions of access to publicly available telecommunications networks and services for both users and new service providers. Harmonization covers three domains: (1) technical interfaces and/or service features, (2) supply and usage conditions, and (3) tariff principles. These conditions of ONP must not restrict access to networks and services.

The EU has approved various directives since 1990 that apply the principles to specific types of network infrastructures such as public-switched telephone networks, public-switched packet data networks, mobile services, intelligent networks, ISDN, and broadband communications networks. For example, the Commission enacted directives on competition in the markets for satellite services and equipment (Commission, 1994a) and for mobile and personal communications (Commission, 1996a). Also, the Council extended the

ONP principles to leased lines in 1992 (Council, 1992) and to voice telephony services in 1995 (Council, 1995) that was replaced by a new Voice Telephony Directive in 1997 (Council, 1997a).

Caby & Steinfield (1994) concluded that these liberalization trends by the EU could be summarized as efforts for

> (a) creating an open and comprehensive market for telecommunications equipment, (b) progressively opening national markets to competitive supply of value-added services, and (c) curtailing the power of existing telecommunications administrations to use their regulatory authority to constrain competition in services or equipment. (p. 42)

Currently, the *White Paper on Growth, Competitiveness, and Employment* (Commission, 1993), with the full political support of the European Council, has placed the EU's telecommunications policy at the heart of the EU's general policy. As follow-up to the *White Paper*, the Bangemann Group's (1994) report on *Europe and the Global Information Society* has confirmed the EU's telecommunications regulatory agenda. The report pleads for a break with the past, ending state-owned monopolies and making rapid progress towards a fully liberalized environment. A strategic approach of this report was delineated in the *Europe's Way to the Information Society* (Commission, 1994b) and is still an underlying theme of the updated *Action Plan* adopted by the Commission in November 1996 and regularly updated.

As Waverman & Sirel (1997) indicated, "while the European Commission recognized the importance of competition in telecommunications, implementing this understanding with long-established state-owned monopolies has not been an easy task" (p. 114). In spite of the EU's harmonizing efforts, each European country has maintained its own unique approach to the provision of telecommunications. Indeed, according to a report in the *Economist*, even if all 12 of the EU countries became a single telecommunications market on January 1, 1998, the actual liberalization would not proceed without a hitch[16] ("In the Shark," 1998, p. 59).

A German Evolution

Under Article 87 I. *l* of the German Basic Law (*Grundgesetz*) of 1949,[17] posts and telecommunications had to be a public administration directly assigned to the federal government. Accordingly, the German market for telecommunications networks was a public monopoly controlled and operated entirely by the state-owned DBP, which ran the network and all the services on it.

The European Commission directives obviously pushed Germany to accelerate liberalization plans. Also, the United Kingdom demonstrated its successful reforms of telecommunications in the 1980s. In light of these, German authorities began to alter their view of a public monopoly system (Waverman & Sirel, 1997, p. 122). Specific change began with the 1989 Law on Restructuring the Postal and Telecommunications Services and the Federal Postal System.[18] The law was mainly implemented to eliminate the structural conflict between the regulatory and the production activities of the DBP. To accomplish this, the 1989 reform divided the business side of the DBP into three sections: the mail service, the post bank, and telecommunications.

The law also separated the entrepreneurial and operational functions of the DBP from its political and regulatory functions. The former are left to be performed by *Deutsche Telekom* (DT), while the latter are the responsibility of the Federal Ministry of Post and Telecommunications (*Bundesministerium für Post und Telecommunikation*, BMPT). Thus, as a member of the federal government, the BMPT carries primary responsibility for ensuring that DT is run according to law. The other objective of the 1989 Law was to initiate the creation of a private market for telecommunications goods and services and encourage competition for any kind of terminal and any kind of service except basic telephone service.

Even though the German reform of 1989 initially resulted in liberalizing telecommunications markets, it did not entirely privatize a previous government monopoly. This interim approach, characterized as "liberalization without privatization,"[19] is due to certain constitutional restrictions. As briefly mentioned above, the monopoly structure and public ownership of DT were made part of the German Basic Law drafted just after World War II. To amend a constitutional provision in the German system, a two-thirds vote in both the *Bundesrat* (the upper house) and the *Bundestag* (the lower house) is required. In the late

1980s and mid-1990s, the ruling coalition forces fell short of the requisite two-thirds. The only way to amend the Basic Law was to convince a majority of the opposition Social Democrats (*Sozialdemokratische Partei Deutschlands*, SPD), who traditionally had strong ties to trade unions, to vote in favor of the amendment. Moreover, the *Bundesrat*, appointed by the governments of the states, was controlled by the SPD (Baird & Stone, 1998, p. 194).

The SPD eventually agreed to a government plan toward more liberalization by signing to amend Article 87 (f) sentence 1 of the Basic Law in 1994,[20] which paved the way for privatization of DT.[21] Under the amended Basic Law, the federal government is merely required to *guarantee* that adequate telecommunications services are broadly available in Germany rather than having to provide them directly. It remains responsible for securing universal service in telecommunications through legislation and regulation.

The culmination of a fundamental restructuring of German telecommunications is represented by the passage of the Telecommunications Act (*Telekommunikationsgesetz*, TKG) in September 1996.[22] This new, overarching law sets the stage for opening the market to virtually all sectors of the German telecommunications industries for private competition. The complete deregulation began on January 1, 1998, in accordance with the European Council's timetable. This new framework also transformed the BMPT into a regulatory authority by designating it the National Regulatory Agency (NRA).[23]

In conclusion, during the 1989-1998 period, the German telecommunications market moved from a model of public provision of infrastructure with an assumption of natural monopoly to a model of privatization, liberalization, and deregulation that is designed to establish commercially competitive markets.

Leading Asia: The Japanese Case

Until recently, the Japanese telecommunications market was controlled for the most part by a legislated monopoly, Nippon Telegraph and Telephone Public Corporation (NTT). It was created in 1952 as a state-owned common carrier for domestic telecommunication.[24] A year later, a private company, *Kokusai Denshin Denwa* Co., Ltd. (KDD), was formed to serve as a monopoly common carrier for international service operations.[25] This made Japan a unique country where the domestic and international common carriage businesses were separate. The monopoly statuses of NTT and KDD were also stipulated in the *Koshu Denki Tsushinho* Law (Public Telecommunications Law).[26] The statute, passed in 1953 by the Diet (parliament), was implemented through the Ministry of Posts and Telecommunications (MPT).[27]

It was not until 1985 that Japanese telecommunications entered the era of deregulation. Even before 1985, however, there were some liberalization moves, as technological development itself made some of the underlying economic rationales for the monopoly system invalid. In 1971, the Public Telecommunications Law was amended to allow the use of leased lines for computer time-sharing connections and for other data applications on a shared-use basis. The 1971 revision also permitted NTT to provide data-processing services (Bruce, Cunard & Director, 1986, pp. 356-358). The 1953 statute was further revised in 1982 to allow the introduction of private value-added networks (VANs) over public phone lines.

The 1985 reform was initiated by the passage of two important telecommunications laws in December 1984. One is the Telecommunications Business Law[28] that stipulated the basic principles and structure regarding the telecommunications business in Japan. The law dismantled the monopolistic structure by introducing competition into both domestic and international telecommunications services. It specifically designated two categories of carriers: type I carriers were facility owners; type II were facility lessors. Whereas type I carriers[29] were regulated by the MPT in entry-exit, service provision, pricing, and other aspects because of their large investment in equipment, type II carriers, subdivided into two categories,[30] were fully liberalized in April 1985.[31]

The other transforming statute is the NTT Law[32] that newly defined the aims and characteristics of NTT. It converted NTT's public monopoly status into a private, joint-stock company through a series of public stock offerings until a maximum of two-thirds of its ownership was in private hands.

In April 1985 when these two new laws went into effect, competition specifically began, as three common carriers were newly allowed to operate nationwide for long-distance telephony, and other new carriers were permitted to operate regionally or with mobile telephones. The sale of telecommunications terminals was also liberalized. Although local telephone markets were left under NTT's monopoly control, international telecommunication was opened to competition in October 1989 when two new common carriers were allowed to enter the international market. As Oniki (1993) summarized, MPT's policy since the 1985 reform can be characterized as permission for new common carriers to operate in an environment more favorable than that in which NTT operated. It is propelled to facilitate competition in long-distance and mobile telephony, to improve NTT's internal efficiency, and to accelerate digitization of its network (p. 70).

In short, the 1985 restructuring drastically changed Japan's traditional telecommunications system by introducing competition into all the arenas of domestic and international telecommunications. As a result of this significant competitive pressure, NTT's and KDD's remaining monopoly has been significantly weakened, even though the two telecommunications firms continue to be dominant in Japan.

However, when NTT was privatized in 1985, the Diet required the MPT to review the organization of NTT within five years. The MPT reviewed it through the Telecommunications Advisory Council (TAC)[33] to determine whether further reorganization of NTT was necessary to create a more competitive market. The TAC issued an *Interim Report on Japan's Telecommunications Industry in the Future* on October 2, 1989. Having realized that effective competition in the telecommunications industry was difficult to obtain because of the monopoly power of NTT (especially in local markets), the report recommended divestiture as a way of dealing with those problems. The MPT, accordingly, recommended the breakup of NTT into a small number of carriers, similar to the breakup of AT&T in the United States in 1990.

The proposal was not implemented, however, for several reasons. For one, NTT claimed the breakup would harm the R&D functions of an integrated NTT and result in rate hikes in rural area. Politicians were concerned about hard feelings of the public toward the government since the NTT privatization. The Ministry of International Trade and Industry (MITI), one of the main actors in telecommunications policy along with the MPT in Japan, also expressed several concerns regarding the share price and further delaying sales to the public (Naoe, 1994, pp. 652-653; Ryan, 1997, pp. 26-27).

Instead, the MPT recommended several measures for NTT to take. In April 1992, NTT introduced an independent system composed of a long-distance communications division and a regional communications division. The two divisions prevent cross-subsidization. NTT spun off its mobile communications businesses by separating them from NTT and named them NTT Mobile Communications Network. They further divided them into nine regional blocks in 1993. Additionally, NTT has facilitated network connection to new common carriers by allowing competitors to access its network at any point a switch exists (Ryan, 1997, p. 26).

Even after the 1990 dispute, the MPT has sought to gain support of its view that NTT needs reorganizing through some kind of divestiture, even though both NTT and the MITI claim that a divestiture would damage NTT's ability to compete in the upcoming global information infrastructure (Nohe, 1996, p. 327). The issue of the breakup of NTT returned to the forefront of debate in 1995, but the idea was again resisted by NTT, the MITI, and the governing coalition for reasons similar to those raised in 1990. Some deregulatory measures were agreed to, however. KDD was allowed to enter into the domestic long-distance market and NTT was required to provide more interconnections between private leased lines and its public network (Ryan, 1997, p. 26).

NTT's expansion of interconnection sites has led to rapid growth of competition in the local call market. Such interconnection was not required by law, but resulted from private negotiations between carriers. However, with the increase in the number of carriers entering the market, it became necessary to establish systematic interconnection rules and therefore the Telecommunications Business Law was revised in June 1997. The major revisions were the requirement of type I carriers to establish interconnections at the request of other carriers and the enforcement of some obligations on carriers operating designated facilities.

Accompanying the clarification of interconnection rules, progress was made toward a more deregulated environment. For example, foreign capital restrictions on all telecommunications carriers other than NTT and KDD were lifted and further efforts are underway to abolish all the restrictions and make KDD a fully privatized company. As of January 1998, the number of telecommunications carriers in Japan was 5,765, of which 149 were facility-based carriers. This clearly indicates that the Japanese telecommunications market is well on its way to becoming competitive.

Theoretical Framework: Political Economy

In submitting telecommunications deregulation to critical investigation in terms of public interest values on the Internet, this study identifies political economy as the theoretical framework that most coherently describes and explains the phenomena being examined. Political economy "defines itself in part against a range of intellectual challenges" in which the policy perspectives flowing from neoclassical economics "have historically embodied the most significant intellectual opposition and continue to do so" (Mosco, 1996, p. 81).

This section explicates various theories of telecommunications regulation and compares their strengths and weaknesses in terms of political economy. By doing so, it mostly relies on Mosco (1996) to provide a historical overview of political economy as a general approach to social analysis, to explain its application to mass communications, and to synthesize its contributions to and debates in communication studies. Mosco has made an ambitious attempt to rethink and renew political economy by exploring its relationship to policy studies and to cultural studies. This section is also attributable to Horwitz (1989) who offered well-summarized explanations of various perspectives on theories of regulation.[34] He also concentrated on the deregulation phenomenon through the history of United States telecommunications regulation and shed substantial light on the complex mosaic of political, economic, legal, and technological forces that have undermined the traditional reasoning for the regulation of telecommunications.

Responses of Political Economy to Theories of Regulation

The identification of major theories of telecommunications regulation differs from scholar to scholar. For example, Brock (1981) classified them into public interest approach, modified public interest approach, economic approach, status quo approach, political approach, and contract approach (pp. 9-11). Horwitz (1989), on the other, identified the following theories of regulation: public interest theory, perverted public interest theory, conspiracy theory, economic capture-conspiracy theory, organizational theory, and capitalist state theory (pp. 22-45). But though some branches have developed that are extensive and sometimes confusing, there are still two basic theories originated that explain the regulation of telecommunications markets: public interest theory and economic theory (Wenders, 1989, p. 105).

Both general theories and policy studies grounded in them take an instrumentalist position that explains the process of policy formation by describing the motives or actions of special-interest groups that lobby the government to act according to their interests. For instance, the essence of the economic theory[35] is that regulation is a crucial mechanism by and through which industries or occupational groups seek protection from competition. Regulation is understood as measures demanded by many industries to limit entry, raise prices, or otherwise reduce output where the private costs of cartelization are too high.[36] The regulatory measures are supplied by politicians who, in return, get political contributions and other favors. Because the industries' potential gains from regulation are enormous, "every industry or occupation that has enough political power to utilize the state will seek to control entry" (Stigler, 1971, p. 5) in order to maximize its profits.

Political economy, instead of taking the instrumentalist explanation, "aims to deeper levels of understanding that start from locating the communication industry within the wider totality of capitalist social relations" (Mosco, 1996, p. 263). As many political economists (Garnham, 1990, p. 7; Golding & Murdock, 1991, p. 18) have pointed out, political economy starts with sets of social relations. It is as interested in seeing particular aspects of social life as are other theories. But what makes political economy distinctive is that "it always goes beyond situated action to show how particular micro contexts are shaped by general economic dynamics and the wider structures they sustain" (Golding & Murdock, 1991, p. 18). Not only is political economy used as a means of understanding the specific facet of society from the social whole, but it seeks to identify relationships that prevail in the entire social arena. Indeed, political economy pays attention to the interplay between economic organization and political, social, and cultural life. In short, as an analysis of the wider social totality, political economy transforms its problematic by asserting that

> understanding *how* requires more than comprehending *who* does *what* to *whom*. Moreover, *how* alone does not explain the process. This is accomplished by determining *why* actions take place, and this demands locating communication policy within the general political economy, including historical and contemporary tendencies. (Mosco, 1996, p. 263, emphasis in original)

A commitment to the social totality specifically requires understanding the relationships between the political and the economic dimensions without losing a sense of fit between them.[37] However, the two mainstream regulation theories tend to give excessive emphasis to either the political or the economic. First, public interest theory exemplifies a statist inclination having an assumption that ultimate decision-making authority resides with the state.[38] Established in response to conflict between private corporations and the public, this theory rests on the capability of state intervention to secure both socially desirable economic redistribution and general economic efficiency. The theory continues to say that regulation is aimed at avoiding both monopolistic abuses by corporations and market failures stemming from externalities[39] and economies of scale. State agencies, in this respect, are set up to curb abusive business practices and promote economic efficiency.[40] In short, a statist approach like public interest theory treats the government as the determining power for policy-making processes. Moreover, this approach simply assumes such good intentions of government in establishing regulatory agencies that it is unable to adequately explain regulatory processes.[41]

Second, the opposite tendency is to work from an economistic view of telecommunications. It has long been argued that regulation is vital for many industries to control entry and construct artificial cartels. This micro-economic standpoint of regulatory approach is typical of almost all theories of regulation. As Mosco (1988) observed, "where telecommunications policy research is analytical rather than descriptive, it has tended to rely excessively on ... economic – typically neoclassical economic – reasons for what propels policy making" (p. 107).[42] The tendency emerges from the critical importance of continuing telecommunications infrastructure innovation and investment to economic development. In particular, the recent accelerated thrust to deregulate the telecommunications industry has brought wider attention to economic rationale for policy formations, as deregulation is the logical outcome of a liberal market philosophy. Adopting a deregulatory policy in the telecommunications sector is an attempt to modernize the economy through privatization and promotion of competition.

These two extreme approaches taken by public interest and economistic theories provide only a partial explanation in that they sacrifice other critical elements of regulation and ignore the interaction between them. Therefore, they fail to offer a complete understanding of various components of telecommunications regulation such as state regulatory institutions, telecommunications industries, and civic groups at either the domestic or international levels.

But political economy situates itself within the broader context of the social system and its relations to other social systems in order to see how a specific structural force influences the other realms of society. The approach is therefore similar to capitalist state theory developed by recent neo-Marxist theorists. The theory positions every actor involved in policy-making within much larger structures of power and constraint in the period of advanced capitalism. Regulatory agencies, then, are "an institutional response of the political system to the demands and contradictions created by the emergence of monopoly capitalism" (Horwitz, 1989, p. 41).

In addition to the concerns about complete elements of social relations, political economy focuses on the structure of social power. Social class is the starting point for the formation of social relations and class power in particular occupies a central place in political economic analysis. According to Mosco (1996), the main concern of political economy is "a specific set of social relations organized around power or the ability to control other people, processes, and things" (p. 25).

Power has two important characteristics in political economy. First, it is built into a social system. Thus, regulatory agencies are not accepted as a given, but must be connected to existing power structures. Second, power is a form of control. As a central force in shaping social relations, it is used to preserve a privileged status against challengers within social hierarchies. In both regards, political economy tends to focus on the ruling class. In particular, it "seeks to explain how that class is constituted through control of the production of the material and cultural filaments of society" (Sussman, 1997, p. 4).

In contrast, public choice theory[43] treats power as diffused and dispersed throughout society. Power is seen as distributed across many competing interests that vie with one another to create a more or less balanced political equilibrium. Public choice theory has created a policy science by interrelating the concepts of the market, individual choice, and private self-interest. It starts from an observation people are pursuers of private self-interest. Individual choices reflecting this private self-interest constitute collectivities, while the market provides the structural model for all collective activity. Normative judgments are based on how well such activity models the maps of perfect markets – the ideal types in economic theory. In reality, individual utility maximization is realized through the choices of participants across a range of imperfect markets that form social life. Social order is transformed in this ongoing process. Public choice theory leads policy studies to conceptualize the normative warrant in order to improve the social order, because this is how people best realize self-interest.

From this public choice perspective, there is no structural tendency for power to be concentrated in one group or interest. Its political tendency tends to be pluralism, which assumes that power is a widely dispersed resource more or less accessible to all interests operating within the political arena (Derthick & Quirk, 1985). Thus, questions of power are rarely raised because power is not considered problematic.

Telecommunications policy generally involves objectives and measures to realize social goals and is concerned about public interest values. These are the focus of normative theory generally.[44] In a way, any theory of regulation is typically normative "in that it seeks to influence public policy formation, improve the functioning of government organizations, or alter the institutional framework within which the process of regulation takes place" (Trebing, 1984, p. 3). Because any normative theory is necessarily based on reference concepts and policy objectives that inevitably incorporate some value elements, "normative approaches have had a considerable influence on the process of policy formation" (Bauer, 1994, p. 29).

Nonetheless, regulation theories have not always had interdependent or complementary relationships with normative theory. In particular, many economists' approaches cannot evaluate normative values, which are not measurable in purely economic terms. Thus, normative comment would hold little or no explicit place in the economists' assessment. From their viewpoint, regulation is a device whereby politicians transfer power to well-organized groups if the groups will return the favor with votes and contributions. As Stigler (1971) admitted, regulatory measures are "acquired by the industry and [are] designed and operated primarily for its benefit" (p. 3). In this vein, regulation is not designed to defend the public interest and does not care much about normative values. Regulation of telecommunications becomes "industry-inspired and profitable" (McCormick, 1989, p. 21) without due consideration of social, political, and public aspects of it.

This lack of normative concern makes economic approaches to regulation of limited value, even if they have been useful in keeping us abreast of developments in technology, services, and changes in the players and their relationships. Economic approaches, for instance, have improved our understanding of the telecommunications industry and contributed to the overhaul of established policies. But political economy calls for a critical "understanding to a set of values or a moral philosophical standpoint that assesses the process for its contribution to democracy, equality, participation, fairness, and justice" (Mosco, 1996, p. 263). Mosco uses moral philosophy to determine social values and conceptualizations of appropriate social practices (p. 34).

From the time of its founders, political economy has been grounded in a utopian outlook that views the role of communication in terms of creating a desirable social environment. It is a vision for a better society in which communication performs a major task. For example, Karl Marx began with moral philosophical treatises, which formed the core of understanding the values of a growing industrial society. One of his theoretical rationales for concern about the moral or normative dimension in political economy comes out of this rationalist notion of truth. Discourse constructed through all mediated forms of communication is capable of demonstrating a truth about a world external to that discourse. It has a determinate relation to the actions of human agents, actions about which it is possible to make normative judgments. In other words, such actions can be meaningfully interpreted as good, bad, or mistaken (Garnham, 1990, p. 5).

Principles of Political Economy

Until now, distinctive, yet interrelated features of political economy have been identified in three main respects. First, political economy is a way of exploring the social whole. Although it starts with particular aspects of social phenomena, they are framed and guided by looking at a broader level of social totality.

Making use of the social totality does not require an essentialist or a reductionist view that all is reducible to the social whole, because dialectical thinking about the relationships among key elements in the wider social arena leads the approach to recognize that reality consists of both parts and a whole (Mosco, 1996, p. 33). Giving importance to the social totality broadens the base of telecommunications policy analysis, because it can then incorporate political, social, and economic actors, and forces that mutually constitute the policy-making process.

Second, social totality is the concrete manifestation of not only mutual interests of major social components, but also structures of power. In other words, all aspects of social reality are "mutually constituted with objective conceptions of power that derive from the fundamental rules governing structures in society" (Mosco, 1996, p. 258). Because political economy places power at the center of its fundamental way of thinking about social relations, it is defined as "the study of the social relations, particularly ... power relations, that mutually constitute the production, distribution, and consumption of resources" (Mosco, 1996, p. 25).

Political economy does not claim that all forms of social power come from economic relations. Nevertheless, it tends to postulate a hierarchy of determinations and a related hierarchy of analytical priorities. Political economy is particularly concerned with the analysis of economic relations from which class relations devolve. Over the last two decades, telecommunications systems have been substantially altered by deregulatory policies in which the market replaces the government. The market is not only a regulator of telecommunications institutions; it is itself a communicator of enormous power. How this power is exercised is of major interest to political economy.

Finally, one of the central subjects of political economy analysis is the normative or moral domain. As Golding & Murdock (1991) maintain, what distinguishes political economy from other theories is that "it goes beyond technical issues of efficiency to engage ... basic moral questions of justice, equity and the public good" (p. 18). As noted above, normative viewpoints are often masked in general regulation theories, especially those based on economics.

By paying attention to normative matters as well as societal power structures, conflicts of interests, and the ideological role of telecommunications in social processes, political economy makes it possible to be concerned with the concrete consequences for the public interest in telecommunications deregulation as established within the broad patterns of power relations. Mosco (1996) expected that "this valuational stance helps to lift policy analysis out of its narrow, descriptive, and positivist tendencies by connecting the analysis of policy to the politics of changing it" (p. 263).

In a practical and specific sense, these three principles of political economy require two key steps in addressing a focal question for this study. One is to chart the shifts in the balance between the market and the government after telecommunications deregulation. The other is to trace its impact on public interest values such as universal service and free speech. Given the first two characteristics of political economy, this study will locate telecommunications deregulation within the wider context of political and economic forces at the global level, describe the current institutional map that constitutes the policy process, and analyze the configuration of the social relations, especially power relations, that currently exist. The third feature of political economy will guide the other task by investigating how deregulatory changes have altered the array of forces that exercise control over telecommunications systems and thereby either limited or liberated the public interest.

Political Economy Research on Telecommunications

Academic interest in telecommunications policy has never been greater owing to a natural outgrowth of the increasing awareness of the importance of telecommunications in modern society. The current mapping of this area was provided by Bauer (1994) who mentioned, "telecommunications research has become a rather complex, truly multi-paradigmatic undertaking" (p. 18). He listed two reasons. For one, many different disciplines such as economics, political science, sociology, law, computer science, engineering, management science, and communications all address telecommunications policy issues.[45] For another, various traditions of thought even within disciplines have made significant contributions to the body of telecommunications policy research.

As a result, policy research in telecommunications has taken many shapes and is now positioned along several different theoretical perspectives. This section deals only with scholarly literature on telecommunications from the viewpoint of political economy. Even so, just the literature on political economy is so vast that a comprehensive review is impossible. This review is intended to demonstrate typical or popular patterns of political economy studies regarding telecommunications rather than to evaluate and generalize all telecommunications-related works from that approach.

One of the major substantive themes in political economy has been to locate the relationships of government, or the state, to the telecommunications business within the wider context of changes in the power relations of telecommunications and new technologies. Danielian's work in 1939 became pioneering research. It described the structure and practices of the largest telephone company, AT&T, and addressed concerns about the use of power by such large enterprises. Smythe (1957), following themes laid out in the Danielian's research, produced early important policy research. As one of the founding figures in this field, he recognized the tensions between the state's general tendency to mediate the interests of leading corporate forces and its potential to serve as a vehicle for publicly controlled services. Mahon (1980) directly inspired political economists to reflect on a fundamental problem facing government in more advanced capitalism by asking how the state could promote the interests of capital as an independent arbiter of the wider public interest.

Their problematic has been further developed by such political economists as Mosco (1982) and Schiller (1982) who emphasized the relationship between government and corporate power. Mosco found that the electronic power structure contains forces comprising established producers of information sources, new producers, and transnational users who seek to shape the state into various molds that reflect their specific needs. These conflicts are embodied in state agencies allied with different capitalist interests as well as the different functions of capital accumulation and legitimation. Schiller's work is important because it identifies large, well-organized business users as a new powerful force in the telematics[46] industry that signaled a principal shift in the industry power structure. As business users demanding advanced telematics services had mustered policy-makers' support effectively, it was expected their private control would be enhanced over not merely information technology – but the economy and society as a whole.

Against this backdrop, the contradiction of the capitalist state was discussed in the area of telecommunications policy by Calabrese & Jung (1992) and in the development of new information technologies by Gandy (1993). Calabrese and Jung examined the impact of market-based policies on rural residents and rural life and assessed the response of movements organized around rural rights. Based on the assumption that the cable and telephone companies' rural development should not get in the way of providing the public with access to a wide range of broadband services, they argued that the state could contribute to the solution by encouraging the design and experimentation of alternative broadband development schemes that would not favor one industry over another.

Gandy (1993), a typical political economist who sees the information age as merely an extension of capitalist influence beyond its traditional industrial base, investigated the capabilities of new technologies to commodify, control, and reconstruct personal information and personal identity. Emphasizing the role of government in the collection or generation of information about individuals and groups in society, he expressed concerns over the limitations on the public access to that information. This concern was heightened by a tendency on the part of government to implement a policy of privatization, which makes public access dependent on one's ability to pay the commercial third-party vendors who are now replacing traditional government information sources.

McChesney (1993) and Winseck (1993) demonstrated an interesting approach by dealing with power relations from a grassroots historical perspective that featured the role of labor and social movement organizations. McChesney, centering on the activities of the broadcast reformers in the early 1930s, showed a good deal of sympathy for their cause. He found the reform movement's critique of commercial broadcasting compelling as was its argument that a democratic society was best served if a significant portion of its broadcasting was not subject to the capital accumulation process. More important, he regarded the immediate aim of the broadcast reform movement – that of establishing a public debate over how to best to organize media services in a self-governing society – to be commendable in principle.

Winseck (1993) examined the history of labor involvement in the development of telecommunications policy in Canada, where profound political and economic changes affected relations between the state, economy, workers, and the public. His dissertation showed while labor unions participated in regulatory proceedings mainly because they realized that policy changes could have greater implications than, for instance, technological change, they also contributed by suggesting new regulatory methods and by supporting basic principles like universal service and public participation.

Some studies conducted in this tradition have been based on the assumption that the dynamics of the telecommunications environment can be appropriately explained by looking at its economic conditions. In this sense, some scholars have further focused on the question of telecommunications business concentration. It was documented in the telegraph, telephone, and new information industries by Du Boff (1980; 1984), Babe (1990), and Mosco & Wasko (1988), respectively. Du Boff is a leading scholar who pointed out that seeds of the interdependence between industry structure and policy making were sown with the first telecommunications breakthrough, the telegraph. He underscored the profound importance of the medium to business growth in the 19th century. His major finding included the acceleration of consolidation among the various telegraph services from 1853 to 1860.

Mosco & Wasko (1988) compiled 14 articles addressing the social relations of information from the viewpoint of political economy. Political economy made it possible to see how power was used to shape the production, distribution, and use of information as a commodity. By doing this, they showed how corporations made information a marketable commodity, how the state advanced that process, what the implications for access to information were, and what the impact was on the workplace and the home.

A slightly different pattern of inquiry addresses the social and political changes in a new telecommunications regime. Hills (1986) produced an analysis of political change in telecommunications that connected a changing policy environment to changes in the relationships among the state, capital, and labor in three leading countries. Hills (1991) followed this with an analysis of the distribution of welfare within the United States and Europe. She argued that research and development, market structuring, and standardization were all facets of a supply-side industrial policy on the part of governments. This was intended to place their economies ahead in the race for international competitive advantage and international power. Such initiatives, however, ignored the complementarity between the use of advanced telecommunications and the diffusion of telecommunications technology. What Hills referred to as "the democracy gap" has been a fundamental theme through the bulk of critical research.

Schiller's (1996) research is worthy of attention in this respect. He contended that there was danger in information inequality, as schools, libraries, and information were subjected to bottom-line measures of the market and made available only to those who could pay the market price. Hi-tech media and the global superhighway, in the hands of multinational corporations, would further reduce the ability of most citizens to make choices between community needs and individual desires. According to him, this unchecked private economic power was a source of the social crisis afflicting the United States.

One of the specific issues related to the impacts of telecommunications policy is explanation of the changes that have taken place with the opening of the telecommunications market. First, Mosco (1989) and Wilson (1992) analyzed the discourse changes that accompanied structural changes by exploring the roots of a shift in the dominant rhetoric from that of public to market-based service. Mosco described deregulation as a metaphor, more appropriately a euphemism, but it did not mean that deregulation was inconsequential. Rather, Mosco argued that deregulation was an instrument to carry out the national and global restructuring of an industry that encompassed culture, media, telecommunications, and information.

Wilson (1992) particularly criticized the very core argument of deregulatory proposals: that deregulation improved the efficiency of the telecommunications industry. By looking at the economic theory that underscored the claims of deregulation proponents, he showed cases where sustainable competition might be inefficient and urged enlarging the arena of debate on telecommunications beyond the parameters created by the market economists.

In a broader manner, both Garnham (1990) and Mansell (1993) drew on a critical reading of public sphere theories and dealt with the threats to public life that grew out of reforming telecommunications along strict market lines to meet the market demands of consumers rather than the needs of citizens. Garnham especially attended to problems involved in the formulation and operation of United Kingdom's telecommunications policy. After examining two contrasting positions in attitudes toward telecommunications deregulation, he argued that competition was unrealizable and inefficient and that a monopoly was both unsustainable and undesirable for technical, economic, and social reasons.

Another important issue for public policy was dealt with by Mansell's (1993) analysis of whether the evolution of advanced public networks would extend access to new services or create new barriers that would limit these services mainly to the most privileged customers. Based on the findings that disparities in network access conditions would continue to arise, he argued that the power of the state should be exercised via the public regulatory process.

As "the global construction of a new information infrastructure is widely recognized as a defining characteristic of the present epoch," (Mueller, 1995, p. 458) political economists have given considerable attention to telecommunications businesses and the state in the international arena. Given the progressive globalization of telecommunications markets, much of their research aims to understand the global structure of power in the telecommunications industry and its consequences for less developed societies.

Gershon (1997) focused on the transnational media corporation that represented a natural evolution beyond the multinational corporation of the 1960s and 1970s. He first explained the regulatory and economic reasons prompting the formation of such transnational companies and then examined how transnational operations affected the development of new media products in terms of cost, quality, and availability. He further provided a series of case study analyses of five leading transnational media companies, giving specific attention to the history, business philosophy, and economic performance of the companies.

In a similar context, Hamelink (1994) used the term, "world communication," to refer to all cross-border traffic of data, information, and knowledge. The major object of his analysis was political practices through which the conduct of actors in world communication was governed. Hamelink assessed the politics of world communication from a human rights perspective because most preambles to regulatory texts were full of people-centered discourse: Are the interests of ordinary people served by the collective practices of world communication politics? He concluded with a proposal for a political practice inspired by the recognition of people's fundamental rights to communicate.

On the other hand, McChesney, Wood & Foster (1998) and Herman & McChesney (1997) inquired about the implications of the current transnational trend in telecommunications for less developed regions. McChesney, Wood & Foster challenged the prevailing belief that corporate transnationalization, presented as "globalization," was a natural and inevitable development. They examined the lower levels of the social pyramid to understand the impact of the process. Markets, profits, and class domination also underpin the computerization of the economy. Those unacknowledged factors were examined in 14 articles mostly grounded in historical experience and analyses of fundamental social relations.

Herman & McChesney (1997) traced the emergence of the global media system (including the Internet and digital communications systems) and chronicled the political, economic, and technological factors that had led to its ascension. Then, by analyzing the implications of the globalization and commercialization process in the United States and several other nations as well, the authors contributed to an understanding of the economic and political dynamics of the newly restructuring global communication environment.

One of the characteristics in the globalization process is the transition from America's postwar hegemony to a world telecommunications order led by transnational businesses and supported by their respective nation-states. This has led to research on the specific characteristics of information businesses by D. Schiller (1994). He also looked backward and forward as well as across the disciplines in addressing both historical and policy concerns at stake in the processes of information commoditization. At issue were successive technologies of information production that, in the context of capitalism, demanded a profound reconceptualization of the term "culture."

Overall, one of the major contributions of political economy research on telecommunications has been to situate it within the broader context of changes in the power relations of politics, economics, and new technologies. Political economy, however, has been inactive in addressing specific telecommunications policy issues such as the impact of telecommunications deregulation on normative or public interest values. Such is the focus of this study. Furthermore, there is no systematic study comparing various liberalization and privatization policies in order to offer the benefits of different views of the national and international developments in telecommunications policies.

CHAPTER 3

Methodology

The primary objective of this chapter is to delimit the scope of, to describe the method of, and to indicate the primary sources for this study. While the theoretical framework of this study – political economy – provides a broad guideline for analysis, this chapter aims to specify practical instructions for the study.

Scope

In identifying and justifying the research problems, this study demonstrates some distinguishing features each addressing a different aspect of research concerns. By listing and discussing them, this section both sets bounds of this study and clarifies what this study is about and for.

First, this study is by nature comparative, examining the experiences of the United States, Germany, and Japan with regard to ensuring public good (universal service and free speech) on the Internet in the wake of telecommunications deregulation in each country. A comparative study, defined as research comparing "two or more nations with respect to some common activity," (Edelstein, 1982, p. 14) is the typical approach to the analysis of telecommunications policy. As telecommunications and other closely related technologies provide the capacities for instant information transfer and communication throughout the world and as the cost of transmission continues to drop and become distance-insensitive, international collaboration or conflict surrounding the development of telecommunications and its impact assumes ever-increasing importance.

Accordingly, one country's domestic telecommunications policy cannot be developed effectively without recognizing its regional and international dimensions. "There will be strong impulses to learn from one another through comparative studies" (Edelstein, 1982, p. 131). This comparative analysis is however conducted only to draw some conclusions about the relationship between telecommunications deregulation and public interest, not to generalize worldwide trends in the area.

Second, this study is normative. It studies public interests on the Internet covering a wide range of social objectives and individual values that have come under threat with the introduction of deregulatory measures in telecommunications. McQuail (1994) said the concern of normative theory leads inevitably into questions of ideology, politics, law, or ethics that are not easy to deal with in a traditional scientific way. This has left normative or ethical matters to specialists other than communication scholars. However, "this state of affairs is unsatisfactory at a time of considerable change and reconstruction of media institutions [including telecommunications systems], when normative questions need to be faced" (p. 123).

This study addresses normative questions by analyzing administrative, judicial, and legislative measures taken to achieve social objectives. In doing so, this study places the researcher into the role of "an active participant in the social process under investigation" (Mosco, 1996, p. 79) rather than a disinterested examiner who seeks only to produce objective findings.

Third, this study is concerned with "impacts" of telecommunications deregulation on public interests regarding the Internet, but the impacts do not refer to the direct, short-term, and intentional effects that most media studies in the tradition of social science have addressed. Indeed, the main variable of this study – telecommunications deregulation – is actually an ongoing historical process begun in the mid-1980s. In contrast, the other main variable of this study – public good concerning the Internet – is a relatively recent subject, even though the Internet was born during the 1960s. Therefore, it is next to impossible to measure the direct effect of the former on the latter. Instead, this study analyzes "impact" in the light of broad relationships between a deregulatory telecommunications atmosphere and public interest value on the Internet.

In the end, this study seeks to contribute to public discussions on telecommunications policy in other countries. The impact of telecommunications on political, social, and economic prosperity depends on corporate strategies that facilitate the development of new telecommunications technologies and expand the range of information services available on a cost effective and universal basis. The effects of telecommunications advances also rely on public policy decisions, as telecommunications development influences not just commerce and consumer equity, but also broader public interest values such as how effectively and equitably all citizens in similar circumstances may be able to exercise their right to govern themselves. This makes the telecommunications policy-making process more complicated and more ambiguous than that in the other price-and-entry industries.

Dye (1998) defined public policies as "whatever governments choose to do or not to do" (p. 2). It is interesting to note that he contended that government *inaction* could have just as great an impact on society as government action. These public policies are better implemented when both broad public discussion of the vision for the future of telecommunications and public participation in telecommunications policy-making process become active. By providing some useful lessons on the relationships between state regulation and public interest issues of Internet policy, this study ultimately aims to encourage the democratic and collective discussion of public values that guide the design and implementation of the future of other telecommunications systems.

Method

According to Harvey (1990), methodology is "the point at which method, theory and epistemology coalesce in an overt way in the process of directly investigating specific instances within the social world" (p. 1). When political economists do research, they usually adopt critical social research techniques. Even though the method is extremely varied, it is based on several key building blocks: abstraction, totality, essence, praxis, ideology, history, and structure.

Abstraction refers to the direction of the approach from the abstract to the concrete. One starts with abstract generalizations to get beneath the surface of appearances. Totality is a view that social phenomena are interrelated and form a social whole. Thus, social phenomena should not be analyzed in isolation. Essence means the fundamental element of an analytic process, and praxis refers to practical reflective activity. To critical social researchers, the purpose of doing research is not just about finding out about the world, it is about changing it. Ideology is defined as a class-based world view whose dominant type is said to be hegemonic. Structure is considered holistically as a complex set of interdependent elements resulting from a particular historical manifestation. Hence, any analysis of structure must be located in the context of a wider historical evolvement.[1]

In applying critical social research to an understanding of the impacts of telecommunications deregulation on public interest values regarding the Internet, the measuring stick is the relationships between normative policy objectives and empirical data concerning them. In other words, the general ideal condition of a communications system such as the Internet as a public cultural space that is open, diverse, and accessible, provides a yardstick against which political economy measures both the telecommunications policy and the performance of existing systems and then formulates alternatives (Golding & Murdock, 1991, p. 22).

Therefore, critical social research seeking to construct or evaluate policies has to produce or rely on both empirical and normative instruments. Empirical instruments must generate the range of alternatives from which a choice could be made, point to the consequences arising from each available line of action, and outline a strategy to attain the preferred outcome. Normative instruments involve weighing, comparing, and ranking differently projected alternatives or outcomes in terms of past experience and assigning priorities to each on the basis of their respective consequences for the conditions of human life (Vaillancourt, 1986, p. 167).

The instruments used in examining policies are neither completely distinct nor inseparably fused. The normative instruments are related to the quality of the empirical data available, and the adequacy of empirical instruments relies upon their capability of fulfilling normative requirements. Both tools are tested through concrete application and are assessed with regard to available alternatives (Vaillancourt, 1986, pp. 167-168).

Given the key building blocks of critical social research, empirical detail must be related to a structural and historical totality. Mosco (1996), pointing out that political economy never opposed empirical investigation, said "empirical results neither arose out of a black box nor spoke to one" (p. 79). Therefore, they should be accounted for by locating social phenomena in their specific historical context, by relating to other phenomena within a prevailing social structure that is maintained through the exercise of political and economic power, and by identifying such power as legitimated through ideology.

Not only should empirical data be interpreted within the wider framework of totality, ideology, history, and structure, but they also should be examined in the light of a critical purpose such as the need to advance public interests. From the outset, critical social research considers "the normative to be pertinent for policymaking" (Vaillancourt, 1986, p. 168) and identifies the researcher as a participant actively involved in the societal evolution in question. Indeed, researchers in this strand are so preoccupied with ethical matters and their role in making and evaluating policies that they acknowledge the legitimate role of the normative in the process of choosing among alternative policies. Besides, normative dimensions are frequently a crucial issue in policy analysis, although it has grown up under the influence of the positivistic method of behavioral sciences (Fischer, 1980, p. 1).

In this study, both empirical and normative instruments are used to compare and evaluate three different countries' policies on universal service and free speech on the Internet after telecommunications deregulation. To do this, it is necessary to identify the points to which those two tools must respond in policy analysis. Dye (1998) presented a useful summary of the complex linkages that operated in the process of policy-making. It is described in Figure 1.

<Figure 1> Studying Public Policy, Its Causes, and Consequences

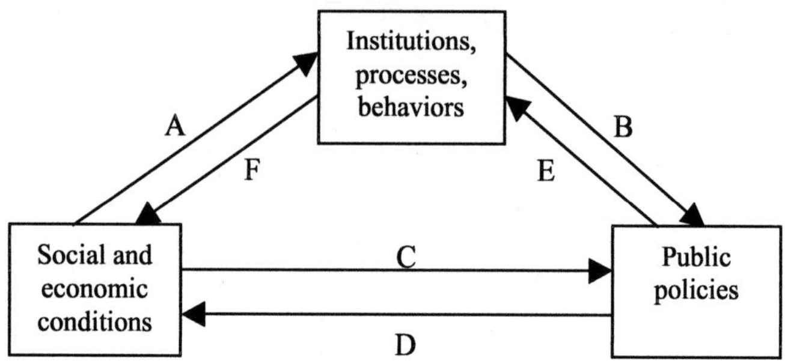

Society	Political System	Public Policy
Including:	*Including:*	*Including:*
Wealth and income	Federalism	Civil rights
Inflation, recession, unemployment	Separation of powers	Educational policies
Educational achievement	Check and balances	Welfare policies
Environment quality	Parties	Health care policies
Poverty	Interest groups	Criminal justice
Racial composition	Voting behavior	Taxation
Religious and ethnic make-up	Bureaucracy	Spending and deficit
Health and longevity	Power structures	Defence policies
Inequality	Congress, president, courts	Regulations

Linkage A: What are the effects of social economic conditions on political and governmental institutions, processes, and behaviors?

Linkage B: What are the effects of political and governmental institutions, processes, and behaviors on public policies?

Linkage C: What are the effects of social and economic conditions on public policies?

Linkage D: What are the effects (feedback) of public policies on social and economic conditions?

Linkage E: What are the effects (feedback) of political and governmental institutions, processes, and behaviors on social and economic conditions?

Linkage F: What are the effects (feedback) of public policies on political and governmental institutions, processes, and behaviors?

Source: Dye (1998), p. 7.

Both telecommunications deregulation and significant public interests concerning the Internet, the two axes of this study, are categorized as public policies in this configuration because they are reflected in and implemented through specific policies. This study does not regard the impacts of telecommunications deregulation on the public good as a matter of inside public policies, however. Rather, the primary concern of this study is represented by arrows B and C because telecommunications deregulation is a phenomenon already evolved. It has significantly affected both categories since the mid-1980s: political system (institutions, processes, and behaviors) and society (social and economic conditions). This is expressed by arrows D and F. The focus of this study is on how universal service and free speech on the Internet (classified as public policy in the configuration) are being ensured by and through relevant public policies within new environments (classified as political system and society in the configuration) shaped by telecommunications deregulation.

Concentrating on this relationship and applying the elementary building blocks of critical social research, this study primarily inquires about the following question: what are the effects of both power structures and social, economic, and historical conditions on universal service and free speech rights regarding the Internet? Nevertheless, this study does not necessarily posit power structures and social, economic, and historical conditions as independent variables and the two public interests on the Internet as dependent variables. Instead, this study is underpinned by a dialectical relationship between them.

Against this backdrop, this study first identifies policy goals and program objectives[2] by locating the enabling legislation, legislative history, staff and committee studies and reports, and by checking other official sources. Enabling legislation refers to acts that contain some statement about the purpose or goal of the policy, and legislative history is a set of official documents or transcripts of legislative hearings and documents accepted as part of the background material for the legislation subsequently passed into law.[3]

But policies need to be assessed by some criteria. Drawing on Fischer (1980, pp. 207-210), this study identifies the following components as policy evaluation questions:

1. Identification of Policy Programs: What are the policy programs derived from the relevant policy goal?

2. Relevance of Policy Programs:
(1) Can the policy programs be grounded in higher principles or established causal knowledge?
(2) Can the policy programs be justified by the political, economic, social, or legal environment?
(3) Is there general agreement about the appropriateness of the policy programs?
(4) Is there any evidence indicating that the policy programs fulfill their stated objectives?

3. Unanticipated, Situational Context:
(1) Are there any circumstances in the situation which require that an exception be made to the policy programs?
(2) Is there any evidence that offsets the program objective(s)?

4. Consistency with Another Policy Goal:
(1) Is there another policy goal equally significant to the situation?
(2) If so, do those multiple goals conflict with each other?
(3) If so, do the policy programs fulfill the objective(s) more efficiently than alternative programs available?

The first policy evaluation enables this study to start with particular aspect of telecommunications policy. This is explored at a larger level of social totality by answering the second and fourth questions. Both interests of social components and power structures in telecommunications, and normative or moral dimensions of telecommunications policy are compelling factors in the examination of the third question.

While empirical instruments have an important role in answering the questions, 2 (3), 2 (4), 3 (2), and 4 (3), normative tools are utilized in the analysis of the other questions. This study does not conduct empirical research itself. Instead, it references empirical data produced by government agencies, international organizations, and other social science researchers.

Sources

Primary sources for this study are both quantitative (empirical data) and qualitative (legislative records, court documents, and archival materials). In addition, congressional debate and committee reports, texts of proposed legislation, and the proposals and commentary of key policy-makers are evidence. The vast majority of the material is in analog form, mostly in print, but has been deposited in digital form on the World Wide Web (WWW) also. This technological breakthrough of converting many analog paper holdings into digital form benefits this study greatly.

Government websites have been most used, although other sites may also provide useful material. Government sites include those of the Federal Communications Commission (FCC), the National Telecommunications and Information Administration (NTIA), the European Union (EU), the European Commission Information Society Project Office (ISPO), the German Federal Ministry of Posts and Telecommunications (*Bundesministerium für Post und Telekommunikation*, BMPT), the German Federal Ministry of Economics (*Bundesministerium für Wirtschaft*, BMWi), the Japanese Ministry of Posts and Telecommunications (MPT), and the Japanese Ministry of International Trade and Industry (MITI).

The FCC, a telecommunications regulatory agency in the United States, has a strong influence on the world's telecommunications policy. FCC documents available at its website have four general types: news releases, orders, Notices of Proposed Rule Making (NPRM), and fact sheets.[4] While news releases are brief, unofficial summaries of FCC orders or other actions, orders are the official statement of FCC actions. Full text copies of all FCC orders since the middle of 1994 are available online. NPRMs are issued when the FCC considers a change to telecommunications regulations and are also available online in the same manner as orders. Additionally, each bureau of the FCC issues fact sheets to present controversial and technical topics, and fact sheets are online as well.

The NTIA, an agency of the United States Department of Commerce, is the Executive Branch's principal voice on domestic and international telecommunications and information technology issues. The NTIA works to spur innovation, encourage competition, help create jobs, and provide consumers with more choices and better quality telecommunications products and services at lower prices. Its website has research reports, filings, and other related material published by the NTIA.[5] NTIA congressional testimony and recent NTIA speeches also are useful for the study.

The EU website[6] is a good starting point for all inquiries on EU telecommunications-related subjects. Almost all information on the whole of EU telecommunications policy is also contained at ISPO's website.[7] ISPO is part of the Information Society Activity Center, which is jointly established by Directorate General Industry (DGIII) and Directorate General Telecommunications, Information Market, and Exploitation of Research (DGXIII) of the European Commission. It aims to promote cooperation and development in the various areas of the information society.

The most useful document for this study is the comprehensive status report on EU telecommunications policy, which is complied and regularly updated by DGXIII. It provides an overall presentation of the current telecommunications policy and includes a description and an analysis of the main policy papers, legislative measures, and promotion activities and is issued monthly. In addition, legislative documents and proposals presenting the main final texts approved since March 1996, the 1998 European Telecommunication Regulatory Package, intermediate documents such as proposals and common positions, as well as a selection of earlier official documents from 1984 to 1996 are available.

The websites of the BMPT, BMWi, MPT, and MITI[8] provide structural outlines of each of these ministries as well as press releases, policy reports published by their own councils or study groups, statistical data, and current topics at issue. All sites are available in both the national language and English.

Websites maintained by government-related research institutes also prove invaluable in identifying the contending issues concerning universal service and free speech on the Internet. Above all, that of the Japanese Institute for Posts and Telecommunications Policy (IPTP)[9] is of great use in this regard. The IPTP was established in June 1988 with the objective of contributing to governmental policy-making by conducting both theoretical research and investigation related to the postal business and telecommunications administration from the medium and long-term point of view. Since the Department of

Economic Research on Telecommunications was established in June 1992, the IPTP has become a think tank for the MPT. Among many items available at its website, *IPTP Monthly Reports* is useful for this study.

Other websites that provide a good deal of relevant information include the World Bank, the International Telecommunication Union (ITU), and the Organization for Economic Cooperation and Development (OECD). The World Bank is a powerful international development agency, headquartered in Washington, D.C. As a brainchild of the Franklin D. Roosevelt Administration, it has long tried to convince less developed countries that telecommunications investments bring economic development. Its website has an *Information for Development Program* (*InfoDev*) including recent quarterly and annual reports, working papers, and news and announcements.[10] *InfoDev* also has a project database that is regularly updated and can be searched either by categories such as country, region, and sector (including Telecom/Policy) or by using keywords.

The ITU and the OECD are the main organizations that track cross-national differences in the telecommunications sector. The ITU, headquartered in Geneva, Switzerland, is an international organization within which governments and the private sector coordinate global telecommunications networks and services. At its website, more than 4,000 titles concerning international telecommunications standards, policy analyses, and statistics are available in printed form.[11]

The OECD was formed by 20 European and North American countries in 1961. Nine countries have since joined the OECD. Currently, the OECD has a somewhat diverse membership, including some less developed countries such as South Korea, Mexico, and Hungary, but it is still largely known as a club for rich countries. Among many items available at its website, *Communications Outlook* provides this study with basic, updated information.

However, the empirical data that are collected from these electronic sources are biased toward economic and technical characteristics. Therefore, this study has utilized many secondary sources in order to supplement the primary materials.

The secondary sources include daily newspapers, periodicals, trade journals, and empirical and critical studies obtained from communications journals, law reviews, and other scholarly publications. Among others, *Computerworld*, the *Economist*, *InfoWorld*, the *Nation*, and *Nieman Reports* provide this study with updated information about telecommunications including the Internet. Indeed, hardly a day goes by without telecommunications events making news reported in one of these publications. Additionally, such academic journals as *Telecommunications Policy, Media, Culture & Society, Journal of Communication, Federal Communications Law Journal*, and so forth serve to identify contending points at issue in relation to the subject of this study and to offer logical evidences to develop arguments of this study.

In sum, the primary materials for this study – policy and legal documents – are collected through the websites of governments, of telecommunications research institutes, and of international organizations, and both empirical and normative instruments are employed to compare Internet policies on public interest values in the United States, Germany, and Japan. This inquiry is designed to reconsider the role of the state in shaping the future of the Internet. In so doing, this study seeks to bring about wide public discussions on the ultimate goals that the Internet around the world will be used to achieve. This study premises that public policies based on this process can establish mechanisms to make sure that all the pieces work together in a desired manner.

CHAPTER 4

Motives for Telecommunications Deregulation

This chapter examines why the United States, Germany, and Japan have initiated deregulatory movements in telecommunications. Given different preexisting national structures and unique cultural, economic, and geographical circumstances in each country, homogeneity is not expected. The reasoning underlying the approaches taken by each country to liberalize its telecommunications sector is no doubt quite different and unique. Therefore, the understanding of deregulation requires explanation – why did it occur in a specific society? At the same time, there are some common, interrelated threads that have made deregulation a worldwide trend in telecommunications.

Common Motives for Telecommunications Deregulation

It is a rare occurrence in history that regulatory measures have been so quickly lifted from a trade sector by so many different countries at roughly the same time. It suggests common factors drive the trend. This section identifies the elements and divides them into political, economic, international, and technological aspects. They are closely interrelated.

Political Considerations

Governmental desire to take advantage of the economic aspects of the telecommunications industry underlies the recent radical structural changes. Since the 1980s, telecommunications has shown good economic growth potential and has become a major driving force for many national economies. This perception of telecommunications is in line with a political acceptance of telecommunications as the foundation of an information-oriented society, as well as a vital infrastructure for developing other industries.

Although the exact nature of the relationship between telecommunications and economic growth relies upon one's perspective on economic development, there is little dispute that telecommunications serves a function in today's economy similar to the railroads, which were vital to American economic growth in the late 19th and early 20th centuries (Twenhafel, Horrigan, Korzick & McCarty, 1989, p. 7). Read & Youtie (1996) specifically identified three ways telecommunications creates economic growth. First, telecommunications creates new firms and jobs. Second, telecommunications permits firms to be more productive and operate in less costly ways. Third, telecommunications services reduce the differences between disparate regional economies by helping link rural communities with urban economic life (pp. 9-16).

To fulfill the promise of economic growth effectively, governments have assessed their own political strength and their ability to generate support within the legislature to determine which regulatory measures would succeed (Martinez, 1992, p. 478). A deregulatory approach is widely believed to be "linked to economic development in that businesses [tend to] be attracted to areas with market-based regulation because of effective and innovative infrastructure, low costs, and more flexibility" (Read & Youtie, 1996, p. 17). Moreover, "there stands an ideological commitment to minimize the role of the state as a direct agency of production, distribution, exchange, and redistribution" (Ure & Vivorakij, 1997, p. 9) in every economic sector after the 1980s.

In short, the change in the governments' perception of telecommunications from a privately or publicly owned monopoly to a profitable industry and the governments' belief in the contributing capabilities of telecommunications to economic development provide part of the explanations for why governments around the world are relinquishing control of the telecommunications industry.

Economic Considerations

Deregulation is the logical outcome of a liberal market philosophy. Recognizing deregulation as central to the broader neo-liberal strategy, Dyson & Humphreys (1989) indicated, "it is part of a grander aim to remove the impediments to the free play of market forces in the domestic economy" (p. 140). Adopting a deregulatory policy in the telecommunications sector is an attempt to modernize the economy through privatization and promotion of competition.

Kahn (1983) provided an explanation of why telecommunications ought to be deregulated for economic growth. In his view, unregulated competition is the appropriate organizational structure for controlling rapid and unpredictable technological change in telecommunications. The benefits of dynamic technological change and risk-taking innovation induced by competition are viewed as outweighing the costs incurred by the duplication of operations. He also noted that increased competition consequent to deregulation leads to a fragmentation of demand and supply as new profitable markets in networks and services attract new competitors who realize latent but unsatisfied demand, anticipate new market niches, and increase the size of the market.

Indeed, the reasoning behind granting increased regulatory flexibility to telecommunications service providers is that private telecommunications firms are more intimately in touch with the needs of their customers and are driven by the profit motive to make the most effective and innovative investments. The most prominent instrument on which a deregulatory model relies is competition.

Competition is extolled as a crucial ingredient. At a simple level, competition among producers reduces costs of production because producers who fail to develop less-costly methods of production are replaced by those who do. Competition reduces transaction costs, because the producer who can match buyers with sellers with the least expenditure of his own resources can charge less for his services. Increased competition leads to market segmentation, which also results in increased product diversity. Keane (1991) clearly summarized some advantages of competition.

> Competition lets individual consumers decide what they want to buy. It keeps prices low and quality high, forces suppliers to take risks and to innovate continually, lest they lose business to rivals offering better, improved products.... Competition ensures freedom of entry into opinion markets for any enterprise which thinks it has something individuals might like to hear, read or watch. Market-led media thereby cater to both mass and minority audiences. (p. 53)

In brief, as Tunstall (1986) mentioned, "deregulation has favored giving competition a chance to show its virtues" (p. 283) benefiting users by providing services with better quality, lower costs, and higher output than they would experience under any form of regulation. A deregulatory approach has been widely enacted for those reasons.

International Considerations

The service that the telecommunications sector provides is, almost as a defining feature, markedly international. The technology of modern telecommunications, more than any factor, has developed a world economy by making the significance of borders and distances separating nation-states less meaningful. With the world market encroaching on every society, there is a growing recognition that each country's domestic telecommunications infrastructure is closely linked to its international competitiveness. Indeed, the ability to compete internationally is the essence of industrial success in a global economy. The telecommunications sector's growing international importance was "already evident to industry specialists by the early 1980s" (Ure & Vivorakij, 1997, p. 9).

State monopolies are major impediments to international transactions in telecommunications services because any policy under the monopoly systems restricts the ability of producers and consumers to interact across borders. Instead, the deregulation of the telecommunications market has become recognized as a decisive step toward making countries attractive locations for business and industry. This is due to the fact that a vibrant, competitive environment in which multiple providers compete with one another is the best prescription for a continuous process of telecommunications product innovation and improvement. Domestic firms' competitiveness also can be improved, as their efficiency and skills are finely honed by domestic competition, which is the key to international success. Thus, telecommunications policies in each country have been evolved to be compatible with worldwide deregulatory market trends and thereby take advantage of new market opportunities at a global level.

Technological Considerations

Technological development has played a key role in fostering competition, at least as important as political, economic, and international factors. Indeed, the use of new, different technologies in transmitting data or information created opportunities to compete with traditional methods. For instance, wireless communications including cellular transmission, wireless cable, and satellite transmissions, are alternative technologies to traditional copper-wired telephony or coaxial cable infrastructures. These alternatives have allowed a substantial number of competing entrants into voice, data, and video transmission business. New technologies such as ISDN and ADSL also allow incumbent operators to upgrade the transmission capacities of their existing infrastructures and lead to the introduction of competition with advanced service providers in the provision of telecommunications goods and services that formerly were subject to a single source of supply. Likewise, new technologies "have fostered this competition, in general adding to the alternatives rather than replacing older ones" (Compaine, 1985, p. 81). In this regard, Perritt (1995) noted, "competition is the best assurance of the benefits of rapid technological change" (p. 95).

One thing that is definitely distinct about technological progress is the speed with which new communication technologies are diffused. The personal computer revolution and the development of the Internet have explicitly shown how rapidly new technologies can be delivered at affordable prices. For example, until 1993 one of the main disadvantages of the Internet as a means of finding and retrieving information had been the absence of user-friendly interfaces that would run on desktop computers. In two years, a variety of Gopher clients and World Wide Web (WWW) clients became available through commercial and non-commercial sources. Information on the web is now accessed through easy-to-use graphical pages that link items through hypertexts, so that any topic is accessible with just a click of the mouse. The adoption rate of the Internet exploded with this development of the WWW, a web of computers within the Internet, which simplified access to a huge volume of information.

As Perritt (1995) mentioned, "it is extremely unlikely that any conceivable legal regime regulating the details of pricing, market entry, product, and service characteristics could have responded fast enough to allow this kind of innovation" (p. 95). Only a deregulatory approach does not inhibit the rapid advance of technological advance. Policy-makers are inclined to adopt deregulatory policies to facilitate this facet of technological development as well as to promote competition in the telecommunications sector.

Individual Motives for Telecommunications Deregulation

The four interrelated factors of politics, economics, internationalization, and technology form the ground from which countries push deregulatory policies into the telecommunications sector. However, the policies must pass through nationally specific conditions before being implemented in each country, which makes every case different from others. This section deals with the concrete, culturally specific motives for telecommunications deregulation in the United States, Germany, and Japan.

Technological Convergence and Free Market Ethos: The American Case

The United States initiated the worldwide deregulatory movement in telecommunications and has provided the world with a successful model of the approach. Deregulatory measures were relatively easy to adopt in a country where "the ingrained free market ethos ... fosters a presumption that all consequences of increased competition ... will be beneficial" (Blumler, 1991, p. 213). According to Zelezny (1997), "experience with suppressive policies of Great Britain during the American colonial period caused many Americans ... to be fearful of any centralized government" (p. 35). Because of this historical recognition of the state as a threat to personal and commercial freedom, the American government's role in determining the competitive environment of the telecommunications sector has been kept to a minimum. Instead, the market is regarded as a neutral regulatory mechanism.

Indeed, such a marketplace of ideas[1] (the conditions under which the United States practices and advocates a free market economy) is believed to be wholly compatible with the American commitment to deregulation in telecommunications policies. Accordingly, deregulatory processes in the United States are mainly based on the economic rationale of dismantling monopolies, which provides sufficient condition for political decisions (Catinat, 1997).

Yet, competition did not come overnight because monopoly had been an industry norm and it was frequently justified by the natural monopoly argument. As Katz (1997) admitted, until the passage of the Telecommunications Act of 1996 (1996 Act), "regulators in the United States, like their counterparts in many other countries, have been hostile to competition on the theory that carriers were in need of protection from competition in order to ensure their ability to invest in new facilities and their willingness to serve all segments of the public" (p. 684). Given this reliance on monopoly, searching for driving forces for American telecommunications deregulation requires an understanding of the motives driving the enactment of the 1996 Act.

The Act represents a vision of a telecommunications marketplace where the innovation of competition completely replaces the heavy hand of regulation. The regulatory model of American media had been grounded in the boundaries between broadcasting and telephony, which was recognized and codified in the Communications Act of 1934. This was due to a distinct separation between conduit and content. With both the digital revolution and converging technologies, however, the once-separate worlds of content and conduit have tended to merge. As they become increasingly indistinguishable, the mental metaphor used to order American communication industries has completely broken down.[2]

According to Krattenmaker (1996), the 1996 Act was designed to address two imperative problems: technological convergence and legal balkanization. The Act specifically attempts to remedy them by: (1) tearing down entry barriers so that legal balkanization no longer stands in the path of technological convergence; (2) changing the mandate of the Federal Communications Commission from deciding who should enter the market to monitoring conditions under which entry takes place in order to control predators; and (3) protecting the most vulnerable from harmful competition (p. 9).

Likewise, it is the convergence of technology that leads to an even further deregulatory trend in United States telecommunications. The tenet that a competitive marketplace is preferable for achieving the purpose has been mostly unquestioned because it is well suited to the political, economic, social, and legal atmosphere in the nation.

Regional Pressures: The German Case

The German approach to deregulation significantly differs from the American one for a variety of historical reasons. In Germany, telecommunications was long believed to be a public resource that necessitated centralized state regulation. Noam (1992) said that the public monopoly system in Germany could be traced back to 1505, when the Hapsburg emperor Maximilian granted exclusive mail-carrying privileges to the Italian Taxis family in exchange for a large share of the profits (p. 3). After that, the state's monopoly was extended to postal services and transporting coach passengers during the 1600s, directly leading to today's giant *Deutsche Bundespost* and railroad monopolies, respectively. In 1892, the comprehensive telephone authority of the state was cemented into law (pp. 10-14).

The public concept of telecommunications fit very smoothly into the larger public philosophy that had characterized Germany during the postwar era as well. According to Hardach (1977), West Germany, since 1948, has been guided by a "socially conscious free market economy," which combined competitive policies with policies designed to correct the perceived shortcomings of market processes (p. 218).

However, several more recent developments have begun to break down the public philosophy. Above all, regional deregulatory trends placed pressure on telecommunications policy in Germany. The European Union (EU) has been decisive in prodding member states, including Germany, in the direction of telecommunications deregulation as well as in setting the pace for it. The European Commission recommended the implementation of harmonious and competitive telecommunications policies as early as 1984, and the 1993 Maastricht Treaty explicitly advocated Europe's economic competitiveness in a world economy. The Commission, acknowledging that the information revolution is as radical a transformation as the industrial revolution, stressed the necessity of efficient telecommunications to enhance the EU's competitive position in relation to North America and Japan in *White Paper on Growth, Competitiveness, and Employment* (Commission, 1993).

According to Catinat (1997), two facts played a key role in the European deregulation. First, the major telecommunications operators came to realize that the maintenance of national monopoly on infrastructure was more and more detrimental to their strategy of economic development. They determined that the statute of national monopoly prevented them from many international alliances they deemed necessary to compete on the global market and to enter new lucrative advanced telecommunications services. Second, the politicians themselves become aware that, in the perspective of an information society, a protected telecommunication sector could only result in higher tariffs and costs for the whole economy.

This awareness was reflected in the Bangemann Group's (1994) report, prepared by a group of notables involved in telecommunications. It specifically recommended that the EU place primary faith in market mechanisms to carry the member states into an information age. As Baird & Stone (1998) mentioned, "the impetus for adopting this policy direction is very strong for a nation [like Germany] that wants to remain a major player in the global economy" (p. 204).

In Germany, the Social Democrats' (*Sozialdemokratische Partei Deutschlands*, SPD) gradual acceptance of privatization of *Deutsche Telekom* (DT) typically represented a radical shift in public attitudes toward the role of telecommunications in a new environment. To be sure, the ensuing deregulatory trend undermines a significant portion of the German rationale regarding telecommunications regulations.

The SPD did not favor the prospect of privatization of DT because such transitions usually entailed massive layoffs, a trend certain to be opposed by trade unions, an ally of the party. Perhaps the most compelling reason for the SPD's new position was the successful British experience in telecommunications privatization as well as the EU initiative. The United Kingdom was an initiator of the government monopoly model of telecommunications. But in 1984, it took the lead in European liberalization by authorizing the privatization of British Telecommunications (BT). This was the first initiative in the Thatcher government's plan to privatize government-run industries during the 1980s. At that time, the government attached great importance to the economic role of telecommunications in a global society. After privatization, the performance of BT greatly improved and other economic sectors benefited from the wide use of the efficient telecommunications system. Now, the British telecommunications market is one of the most competitive in the world.

This British experience itself has become a compelling argument that deregulation and competition enhance the competitiveness of the telecommunications industry and meet consumer needs efficiently. As a consequence, a new public attitude toward telecommunications and its deregulation has taken hold in Germany to the extent that the SPD came around to the position that telecommunications was central to the competitiveness of the nation's industrial sectors, DT was inefficient, employment would be greater with competition than without it, and a privatized structure would provide a wider array of more affordable services (Baird & Stone, 1998, pp. 205-206).

In short, the success of the British experiment in the privatization of telecommunications and the ensuing advent of a new perception of telecommunications were the driving forces for the deregulatory movement in Germany. These realizations made the 1994 reform possible. The legal prerequisites for such a movement were later established with the passage of the *Telekommunikationsgesetz* as a way of preparing Germany for full deregulation of telecommunications services in 1998.

Government Guidance for an Information Society: The Japanese Case

The most important source of pressure for telecommunications deregulation in Japan has been a governmental endeavor to enter into an "information-oriented society." The slogan was first heralded as an inevitable next stage in Japan's economic and social development in the 1980s (Ito, 1986, pp. 215-217). The Japanese government still advocates reformation of every sector to align its society for the rigors of the information society.[3]

The existing monopoly system in telecommunications was seen as an impediment to the developmental objective, and radical structural reforms such as deregulation were prioritized so as to revitalize the economy. Thus, the deregulatory tendency has been significantly boosted, and Japanese industry has expanded itself into information sector businesses as part of the process of "informationization" of society.[4] Even though competition is introduced in accordance with deregulation, "such competition is allowed only with the boundaries set by bureaucrats. Indeed competition is always controlled and coordinated by bureaucrats in Japan" (Matsui, 1994, p. 229). In brief, governmental leadership and guidance have played a key role in telecommunications deregulation in Japan. This is not a unique story at all. The same approach made the postwar Japanese economic miracle possible and transformed Japan into the world's second largest economy.

The two most significant government ministries involved in telecommunications policy are the Ministry of International Trade and Industry (MITI) and the Ministry of Posts and Telecommunications (MPT). The MITI, best known as the architect of the export-oriented growth of electronics in the three decades following World War II, determined that telecommunications would be a key part of the overall electronics industry and a major factor for strengthening that sector. It recently took responsibility for the fact that Japan was lagging behind other advanced countries in developing an information infrastructure and the associated multimedia technology. The MPT, responsible for postal functions and telecommunications and broadcasting regulatory policy, holds the view that there is a social use for telecommunications and it will be assisted by growth of the information society.

The two ministries have generally had a contentious relationship with each other, more than the typical bureaucratic battle over territory. The rivalry comes mainly from their different views on the role of telecommunications in the future information economy. Whereas the MPT regards information service industries as an important object of "telecommunications policy," the MITI sees them as a vital element of "industrial policy."[5] As Leopold (1996) implied, the ministries' feuding sometimes becomes the major stumbling block to deregulation and greater competition (p. 18). West, Dedrick & Kraemer (1997) also said, "many Japanese feel that these turf wars [between the two ministries] jeopardize the nation's multimedia future" (p. 87).

Despite these different standpoints, however, the Japanese political climate is always characterized by "consensus politics" (Ure & Vivorakij, 1997, p. 17). Indeed, the balance of forces between the two ministries has eventually supported structural change in the telecommunications market via a deregulatory approach. In this respect, Jussawalla (1995) maintained, "Japan's policy was the outcome of two conflicting views [of the ministries] on the role of telecommunications in the macroeconomy" (p. 166).

Nippon Telegraph and Telephone Public Corporation's (NTT) monopoly position, for example, was in question in the mid-1980s, as NTT's high rates and lack of innovation showed how inefficient a protected monopoly could be. NTT itself wanted to become more independent of government control. Other private firms also wanted to enter the telecommunications industry, especially makers of telecommunications equipment. They felt that deregulation would strengthen the market and increase their sales. At that time, both the MPT and the MITI agreed that by selling NTT shares, the government would reduce deficits arising from support of Japan National Railways in previous years. Both ministries' alliance for governmental savings became the impetus for the 1985 deregulatory reform, leading to the privatization of NTT (Ryan, 1997, p. 22).

Likewise, mutual consent reached by two major governmental actors in telecommunications policy, the MITI and the MPT, has largely directed Japanese telecommunications toward deregulation, although the rivalry between them permeates the telecommunications policy debate. Because of this consensus politics, the Japanese government can lead and guide even the deregulatory process of telecommunications.

Summary

A set of factors – (1) governments' changing perception of telecommunications as a profit-producing business with an efficient management structure, (2) advent of telecommunications as a major strategic sector for economic development, (3) importance of telecommunications industry competitiveness at a global level, and (4) contributing power of technological development to competition and swiftness of technological progress – has provided fertile soil for worldwide deregulatory movements in telecommunications.

Despite the popularity of the deregulatory policy in the field of telecommunications, the policy transition has had to contend with existing cultures, traditions, and laws of the relevant nations. These nation-specific considerations often take precedence over the four shared factors in the saga of telecommunications deregulation.

The United States has had little difficulty in adopting deregulatory measures in telecommunications. Its prevailing social and political culture has an explicit deregulatory intent. Governmental restrictions are kept to a minimum to enhance a competitive atmosphere. In particular, the 1996 Act was intended to deal with legal problems arising out of the rapid technological development of new information services and to bring the American legal system into line with the requirements of the changing environment.

This was not the case in Germany, where a belief in the public telecommunications system imputed a different meaning to regulation. There, telecommunications deregulation is better understood in terms of regional pressures than in terms of a faith in market mechanisms. EU's initiatives for the economic integration of Europe are the key regional factors for Germany adopting deregulatory policies in telecommunications, as well as the success, for example, of the British experiment in telecommunications privatization.

The Japanese case provides a rather different story. The government's desire to create an information society has been a major force in the reform of the telecommunications sector. Recognizing telecommunications as an expanding economic sector with potential for growth, the government has made efforts to enhance the industry competitiveness by replacing regulation with deregulation; permitting and facilitating intensified competition among telecommunications companies. In particular, both the MITI and the MPT thrust to encourage entrepreneurs in the telecommunications field to develop themselves into viable, profitable industries, even though the two ministries have a feud with each other over control of the telecommunications market.

CHAPTER 5

Universal Service on the Internet After Telecommunications Deregulation

The main objective of this chapter is to examine the experiences of the United States, Germany, and Japan with respect to realizing Internet universal service obligations within deregulated markets. Before dealing directly with this issue, however, the chapter first asks whether universal service on the Internet is an essential component for modern democratic societies. This is largely analyzed by discussing how the Internet has developed and works in a modern society.

Significance of Universal Service on the Internet

Universal telecommunications service is a public policy goal aimed at spreading access to telecommunications to most members of society. The funds necessary to extend hardware connections to individuals are provided by direct or indirect government activity. In the past, universal service was usually accomplished through the principle of the common carriers, defined as entities that provided essential services in a non-discriminatory manner. In exchange for providing such services the carrier received a limitation on its liabilities, as benefiting for an entity that could not discriminate among its customers. Non-common carriers were not burdened by universal service requirements. Instead, the common carriers' profits were used to support some of its end-users, especially residential and rural customers.

The traditional concept of universal service centering on a common carriage system is no longer meaningful. The convergence of computing, communications, and information technologies has generated a much different path from that developed by basic telephony. Additionally, competitive inroads into most segments of telecommunications have limited the ability to generate the funds for such internal cross-subsidies.

Today, the most important universal service issue has become the access to and use of new information infrastructures. The Internet is the most crucial form of communications media with these infrastructures. The fact that universal service obligations exist in every developed country implies that similar benefits from widespread interconnectivity are perceived around the world.

What are the factors leading to such similarity? Perhaps the best way to answer the question is to see how some of society's essential functions relocate within the new infrastructure. Sawhney (1994) stressed the importance of universal service at three distinctive levels: basic individual rights, social system needs, and service to humanity (pp. 378-383). These levels are closely related to prospects for electronic democracy.

The first level is individual. The argument at this level starts with the assumption that access to the many modern communication services of the Internet is a basic human right – not unlike the right to speak. The moral basis of this argument is that every person has a right to these services by the mere virtue of being a citizen.

Ironically, the initial form of the Internet, the United States Department of Defense (DOD) Advanced Research Projects Agency (ARPA)'s network, was not originally intended to provide any access to the general public. It was reserved strictly for military use. The turning point was the advent of USENET NEWS, which resulted from the effort of computer scientists, responsible for the construction of the ARPAnet but excluded from connecting to the defense establishment. They built their own network to provide themselves with communication channels similar to ARPA's computer network in 1976 and interconnected the two networks in 1979. The advent of USENET marked the beginning of a rapid growth in the life of publicly accessible computer networks as well as the end of the military's control of the computer network (Giese, 1996, pp. 129-132).

The ARPAnet split into military and civilian components in 1983, with the civilian component giving rise to what is now known as the Internet. In 1985, the National Science Foundation (NSF) began a program to establish Internet access across the United States by creating a research "backbone"[1] called the NSFnet. For more than a decade the NSFnet opened doors to all educational facilities, academic researchers, government agencies, and international research organizations.

Today, the Internet use is no longer restricted to this elite group. The equipment itself is relatively low cost, connections to the Internet are easily made, and the medium is widely available in public areas such as state universities and public libraries. Hence, the Internet provides access, at least, to a minimum set of information resources and enables more people to have easier access to a greater amount of information.

Furthermore, in a new electronic environment, people are blessed with the benefits of the Internet as much as they are with those of the traditional public sphere such as streets and parks in Western civilization. Just as discrimination, based on such factors as race, religion, and economic status, is not permitted in a traditional public forum, "in an electronic forum, there can be no discrimination because [those] factors ... do not accompany the message" (Taviss, 1992, p. 795). In this regard, Branscomb (1995) maintained that the Internet removes "many of the prejudices that arise from [the] culturally specific visual cues," because it can "ignore space, time, and physical circumstances" (p. 1640). Therefore, an evolving notion of universal service now asks how to secure access to an information infrastructure which represents "the opportunity for all to participate in the main discourses of society; and so stands well within the spirit of democratic participation" (Schement, Pressman & Povich, 1995).

The second level is at the social system. The basic argument is that the provision of services on a universal basis makes it possible for the social system as a whole to function more efficiently. The most significant social change resulting from the Internet is the advent of "virtual community," defined as "a social collective, or community, that forms in the online, electronic world of computer communications known as cyberspace" (Pavlik, 1996, p. 311). Virtual community has a profoundly different shape from yesterday's community in that virtual community, unbounded by geographical, temporal, or other physical barriers, exists in an ethereal electronic space characterized by interactivity.

In this vein, Sawhney (1996) described the Internet as a "fetch" technology. This "fetch" characteristic (as opposed to "deliver") implies that the Internet "would give the users a much broader choice than simple prepackaged options by providing the basic ingredients which they would then mix and match to create customized solutions for themselves" (p. 310). To take full advantage of this strength of virtual community, it is in the interest of society that all members are connected to each other via the Internet. Therefore, no one should be denied the opportunity to participate directly in decision-making and to interact with each other in a cheap and accessible way.

The third level is general humanity. The discourse at this level is based on the natural connection between the craving for ubiquity and the desire for a communion. The Internet is not simply a conduit for transporting information. The connections it creates have more to do with human interaction than do traditional, one-way electronic communications like television and radio. Indeed, one of Internet's potential benefits most frequently touted by grassroots activists is its ability to foster community. United States Vice President Gore (1993), the Clinton administration's most ardent supporter of the information superhighway, noted the new electronic environment is "not merely technology, but communications. Because from communications come community." He specifically cited the Internet as a contributor to the community as well as an enhancer of the quality of life itself. Because the Internet functions as a symbolic thread that ties all citizens together and makes communion possible, the larger the number of humans reached, the greater is the scope of the communion.

For these reasons, the Internet becomes more valuable to people as it connects more people. Universal service on the Internet is recognized by nearly all societies as being of immense social value. Moreover, because the Internet delivers value only to people using it regularly, a universal service policy regarding the medium is becoming worthier of attention. For example, the policy of the European Union (EU) towards an information society has taken into account the need to avoid a "two-tier society," divided between those who have access to the new possibilities and are comfortable using them, and those who are excluded from fully enjoying their benefits (Commission, 1996d). Without universal service, the Internet is more likely to become a communications network for the affluent segments of society, supported by advertising and access fees that might be unaffordable for a large segment of the population.

Three Experiences of Universal Service on the Internet

No precise definition of universal service has ever been agreed upon among policy-makers and researchers. Any definition poses as many controversies as it settles issues. Resolving this ambiguity is not an easy task; nor is it the purpose of this study. Rather, this study adopts a broad view of universal service that probably fits the different experiences of the three nations in question.

Today, universal service is an umbrella term that includes a wide range of policies responding to economic and social needs. S. Miller (1996) described universal service as having five components: access, usability, training, purpose, and affordability[2] (pp. 181-182). This study mostly concentrates on the controversies about access and affordability among a wide variety of subjects now in the headlines, because those two elements constitute concrete examples of the importance attached to the building of future democracies.

Rough Road in Utilizing Commercial Forces: The American Case

In the United States, a universal service principle grew out of 19th century populist concerns about equal access to basic services such as rail transport for agricultural foods. The concept developed as one of the basic themes of telephony policy. Theodore Vail, the architect of the American Telephone and Telegraph (AT&T) system, expanded the concept of universal service in the early 1900s by defining it as a telephone in every home connected to every other telephone in the country (Temin, 1987, p. 16). He had attempted to deliver telephone services everywhere at an affordable cost as a corporate strategy of extending monopoly status for the powerful Bell company in return for embracing state regulation.

The concept of universal service was loosely articulated in Section 151 of the Communications Act of 1934, "to make available ... to all the people of the United States a rapid, efficient, nation-wide, and world-wide wire and radio communication service with adequate facilities at reasonable charges." This notion has evolved with ongoing changes in technology and the marketplace.[3]

Identification of Policy Programs

The most recent codification of universal service was the Telecommunications Act of 1996 (1996 Act) in which the Federal Communications Commission (FCC) identified universal service as one of three main goals. But the Act did not specifically provide the definition of universal service. Instead, it required the FCC to create the Federal-State Joint Board on Universal Service (Joint Board) for making recommendations to the FCC on a formal definition.[4]

In 1997, the Joint Board proposed new rules to implement the provisions of universal service of the 1996 Act guaranteeing affordable telecommunications services, and the FCC adopted them on May 7, 1997. The new rules are grounded in seven general principles: (1) quality and rates; (2) access to advanced services; (3) access in rural and high-cost areas; (4) equitable and non-discriminatory contributions by service providers; (5) specific and predictable support mechanisms; (6) access to advanced services for schools, health care facilities, and libraries; and (7) competitive neutrality.[5]

Among others, the term, "affordability," had not been used before the 1996 Act, but the first principle of the new rules introduces the concept directly and explicitly into national telecommunications policy. It stipulates that quality services should be available at just, reasonable, and affordable rates. Congress left it to the Joint Board and the FCC to determine what an affordable rate was. They found that telephone subscribership dropped as the cost for local service rose above 0.7 percent of annual household income (Cooper, 1996). To ensure local telephone service remains affordable, the FCC has directed state commissions to monitor rates and non-rate factors, such as subscribership levels, the geographic and population size of local calling areas, cost of living factors, and the number of toll calls customers had to make to reach essential services.

Section 254 of the 1996 Act expands the concept of universal service. According to the section, all telecommunications carriers that provide interstate service are obliged to contribute, "on an equitable and nondiscriminatory basis, to the specific, predictable, and sufficient mechanisms established by the [FCC] to preserve and advance universal service." Where universal service is not provided in a community, the FCC or the state may select a particular common carrier to provide such services. That carrier is referred to as an "eligible telecommunications carrier" under Section 214 (e). It has the right to demand that incumbent local exchange carriers share their infrastructure, technology, information, and facilities with requesting small universal services providers in its area (Section 259 (a), 1996 Act). Section 254 also mandates that eligible carriers receive support payments only to preserve and advance universal service. This is consistent with the general requirement that carriers cannot use non-competitive services to subsidize services subject to competition.

The actual mechanism for preserving universal service directly depends on a universal service fund. Traditionally, such a fund has been raised by two types of fees: line charges and usage charges.[6] When the amount collected from line charges fails to recover the entire 25 percent allocated to the long-distance market, local exchange carriers are permitted to recover the remaining portion of their fixed common network costs by charging long-distance carriers for every minute of usage (West, 1996, pp. 171-174).

When establishing the new universal service fund system in May 1997, the FCC cut long-distance access charges and raised the cost of business phone lines and multiple home lines. Those increased costs were to be funneled into the universal service fund and used to connect poorer schools and libraries to the Internet. The FCC order also requires all carriers that provide interstate telecommunication service and pay-phone aggregators to contribute to the fund. Accordingly, local and long-distance telephone companies and providers of enhanced services such as the Internet are obligated to pay into the fund.

In addition, the FCC has adopted a contribution assessment method that is competitively neutral and has appointed the National Exchange Carriers Association as the administrator of the subsidy mechanism.

Overall, the American attempt to fulfill the commitment to universal service is represented by the basic theme of the 1996 Act: the creation of competition across the full telecommunications field, even in areas that had previously been monopoly controlled. Even though this approach tends to increase the possibility that the main actors will be even larger companies than those currently dominating the scene, it is propelled by the governmental belief that a new marketplace will create not only the advantages of competition, but the unforeseeable benefits that result from new synergetic relationships between previously separated businesses and technologies (Meyerson, 1997, p. 287).

Consistency with Another Policy Goal

Using similar logic, the United States government launched the National Information Infrastructure (NII) initiative on September 15, 1993, with the publication of *The National Information Infrastructure: Agenda for Action.*[7] This was published by the Information Infrastructure Task Force (IITF) that was formed by the government specifically to articulate and implement the vision for the NII. Since then, the NII has become a centerpiece of the current Clinton Administration's information policy.

The phrase "information infrastructure" has an expansive meaning. Beyond the physical components of the infrastructure, the value of the infrastructure depends on the quality of its other elements to allow users to access, manipulate, organize, and digest the proliferating mass of information. In the general context of the NII, universal service has been referred to as a process of eliminating barriers so that everyone has the opportunity to use telecommunications systems for effective participation in all aspects of society.

A central premise behind the NII universal service policy is the encouragement of private investment, which ultimately promotes competition (Perritt, 1995, p. 95). In this context, the NSFnet (which began as a DOD communication tool in the 1970s and functioned as backbone infrastructure) was shut down in 1994 when its governmental subsidy was removed.[8] The government has gradually transferred the United States portion of the global computer network to private companies since then.

The major reasoning behind the government's grander plan to deregulate cyberspace was well expressed in the *Agenda for Action* (IITF, 1993). Identifying nine specific principles to guide governmental action, it presented the need for promoting private sector investment to extend universal service to guarantee that information resources were available to all at affordable prices:

> Devising and attaining a new goal for expanded universal service is consistent with efforts to spur infrastructure development by increasing competition in communications and information markets. ... competition can make low cost, high quality services and equipment widely available. Policies promoting greater competition in combination with targeted support for disadvantaged users or especially high cost or rural areas would advance both rapid infrastructure modernization and expanded universal service.

Working under the premise that deregulation would bring increased services, lower costs, and technological development, the government has attempted to realize the policy goal of universal service as well as build the NII within a market context.

Such a goal is consistent with the theme of deregulation that has guided United States telecommunications policy on which the present Internet policy for universal service is also predicated.

Relevance of Policy Programs

In the United States, the goal of universal service is currently being pursued by establishing a wholly deregulated telecommunications market that encourages unfettered private investment. The program can be easily justified within the American social, economic, and political environment where market efficiency has long been regarded as central to the pursuit of the goals advocated for telecommunications policy, and technology itself has forced telecommunications in the direction of increasingly competitive markets.

Many commentators, especially Gasman (1998), have cited the progress of the universal service program as reflected by the fact that Americans have increasingly embraced the information age by obtaining electronic access in their homes. Indeed, the most recent nationwide survey[9] conducted by the National Telecommunications and Information Administration (NTIA, 1999) supported such observations. It showed Americans were more connected than ever before. Access to personal computers (PCs) and the Internet has soared for people in all demographic groups and geographic locations.

These national penetration rates hide important regional, racial, and income differences, however. According to the same report by the

NTIA (1999), significant disparities have been found between Whites, Asians/Pacific Islanders, those with higher incomes, and those more educated and those who are younger, those with lower incomes and education levels, certain minorities, and those in rural areas. Specifically, urban households with incomes of $75,000 and higher were more than 20 times more likely to have access to the Internet than rural households at the lowest income levels, and more than nine times as likely to have a PC at home. Whites are more likely to have access to the Internet from home than Blacks or Hispanics have from any location. Black and Hispanic households are approximately one-third as likely to have home Internet access as households of Asian/Pacific Islander descent, and roughly two-fifths as likely as White households are. Regardless of income level, Americans living in rural areas are lagging behind in Internet access. Indeed, at the lowest income levels, those in urban areas are more than twice as likely to have Internet access than those earning the same income in rural areas.

Even for some groups, the disparity has widened as the information haves outpace the have-nots in gaining access to electronic resources. For example, the gaps between White and Hispanic households, and between White and Black households, were approximately five percentage points larger than they were one year earlier. The division based on education and income level has also increased in 1998 alone. Between 1997 and 1998, the divide between those at the highest and lowest education levels increased 25 percent, and the divide between those at the highest and lowest income levels grew 29 percent. In sum, there is now an even greater disparity in access to the Internet between certain groups of Americans, and pockets of have-nots are mostly composed of those with low-incomes and minorities.

One of the major reasons for this increase in the digital divide can be explained by looking at the actual telecommunications market trends. In the United States, as the private sector has gained primacy, the contemporary marketplace has been characterized by a "rash of mergers" in which companies seek "the synergy that association with the newly deregulated telecommunications companies would allow" (Golding 1996, 74). Since the 1996 Act particularly lifted many telecommunications ownership restrictions, the number of potential competitors has been reduced to a mere handful.

For example, the two mergers between Pacific Telesis and SBC Communications (SBC) and between Bell Atlantic and Nynex in April 1996 shrank the number of regional Baby Bells from seven to five. In July 1998, SBC and Ameritech and Bell Atlantic and General Telephone and Electronics (GTE) announced mergers. SBC-Ameritech served 13 states at the time and was expected to control about one-third of the American 178 million access lines after the merger. Bell Atlantic-GTE would control about one-third of the local telephone market by combining the local market strength of a Baby Bell with the breadth of services of a long-distance and Internet carrier.

Additionally, during the single month of July 1998, AT&T had been on an acquisition spree by announcing three deals to make a joint venture with British Telecommunications to provide global network facilities and communication services, to buy Teleport Communications which provided local telephone service to businesses, and to merge with Tele-Communications, Inc. (TCI), a huge cable television operator.

WorldCom's completion of a $37 billion acquisition of Microwave Communications, Inc. (MCI) in September 1998 has caused many concerns because this would make long-distance telephone service less competitive.[10] A single telephone company would control over 40 to 60 percent of Internet backbone traffic. Furthermore, MCI WorldCom had proposed $120 billion merger with Sprint that could rival AT&T in size. The merged company, as originally planned, would combine the second and third largest American long-distance companies with one of the top five national wireless phone companies and control a majority of the Internet backbone traffic. But the pending deal to create a mega-carrier was collapsed at the hands of regulators in the United States and Europe in July 2000 who were concerned the merger would pose an unacceptable threat to competition in telecommunications.

Today, almost all industry analysts anticipate even more consolidation in the next few years. MCI President Gerald Taylor, among others, was known to predict that the probable outcome would be "four to six global gangs" dominating the world telecommunications market (McChesney, 1997, p. 22).

This consolidation trend is accelerated by the very structure of the Internet. While large institutions and corporations generally buy their Internet access directly from the large Internet Service Providers (ISPs), smaller businesses and individuals purchase their access from local ISPs that lease phone lines and sell access time to their customers.

But as demand increases at rapid rates and competition becomes more vigorous, most local providers will be driven out in the next few years (Baran, 1996, p. 63).

Some larger ISPs have already discontinued subsidies for smaller ISPs to boost their own network capacity. WorldCom had been accused of engaging in discrimination against new Internet entrants by denying them the ability to enter into important "peering arrangements"[11] and forcing new anti-competitive terms on those who did peer with it. In 1997, WorldCom subsidiary UUnet Technologies, which had 50,000 corporate customers, charged about 12 small service providers new monthly rates for T1 and T3 connections that ranged from $2,000 to $6,000, respectively (Girard, 1997b, p. 2). To be sure, this anti-competitive practice arises out of a desire to eliminate smaller ISPs that compete with WorldCom. A move of this kind would cause many small ISPs to go out of business, unless they survive subsequently increasing fees. Even the few survivor providers are likely to lose their competitive power, because they will have no choice but to pass higher charges on to their subscribers.

The most direct result of the market consolidation arising out of deregulatory economic logic is not increasing competition, but the monopolization of the means of communications. A marketplace report from a Boston-based management strategy firm, Atlantic-ACM, lends support to this thesis. It examined transactions over the past five years involving local, regional, and long-distance telecommunications companies and ISPs and concluded that "the result of consolidation fever had been decreased competition, despite the fact that deregulation was intended to increase competition." The report also said that mergers are "not the best thing for consumers" because they "may at some point force up prices" (Weil, 1998). Likewise, because a fully competitive telecommunications market is not necessarily a natural outcome of deregulation, it is hard to conclude that the American universal service program chiefly relying on market mechanisms is appropriately devised to obtain the universal service goal.

As for the prospect of affordability, it is reflected in the seven principles of the new rules proposed by the Joint Board and adopted by the FCC as described above. Even though the mechanism for affordability is based on a correlation between subscribership and affordability and consideration of how much of a burden a reasonable level of telephone consumption places on the household budget, it does not suffice for fulfilling universal service obligations concerning the Internet.

With the government's elimination of Internet subsidies and the commercialization of the Internet, institutions and private individuals seeking to use the Internet provide full support for the Internet through access fees based largely on their level of usage. A growing number of commercial online services provide an Internet connection as part of their service. Because the Internet uses existing telephone lines, users pay only local telecommunications charges for accessing the Internet, regardless of whether their messages go across town or across the planet, and regardless of the amount of data transmitted in their messages. It is likely that such a fee structure will result in higher prices to individual customers than to volume buyers, making affordability a more complex matter. As S. Miller (1996) pointed out, this is "the exact opposite of what occurs when a common carrier standardizes pricing for all customers" (p. 191). In brief, the American method for affordability is inadequate in that its test is simply consumers' ability or willingness to pay, rather than the full participation of the citizenry in the Internet.

Unanticipated, Situational Context

The introduction of a deregulatory policy is also understood as an effort to accelerate technological improvement. However, E. Miller (1996) asserted that technological advance was not necessarily promoted by deregulatory measures. On the contrary, what is developed by deregulation in the telecommunications industry is "an increase in the market power of established firms and in their ability to offer benefits to well-placed customers" (p. 732). The large ISPs' recent plan to boost their networks is completely intended to cater to more wealthy corporate customers.

One of the reasons deregulation is not necessarily linked to technological progress was explained by survey results on 400 local exchange regulators and officials, conducted by the financial services firm, Deloitte and Touche. The survey showed that since the 1996 Act enabling local and long-distance companies to compete in each other's market along with wireless, cable, and utility companies, local carriers have spent more money on marketing campaigns than on new technology (Girard, 1997a, p. 4).

In the meantime, the FCC order requiring the collection of a universal service fund has run into a roadblock set by private telecommunications companies. As soon as the FCC decided the fund structure, commercial Internet providers filed a lawsuit with the courts challenging the federal decision (Solomon, 1997). The fund was to be paid by telcos to support telecommunications services, including the Internet, for rural areas, schools, and libraries. ISPs do not pay into the fund, although they directly benefit as more users are able to sign up for online services (Weil, 1998). This draws ire from telcos, which are lobbying for a change in the fund structure.

Even though the universal service fund is essential for equal-opportunity Internet access, commercial forces downplay the long-term values not associated with profit maximization and are making it difficult, if not impossible, for universal service to be realized.

Market Mechanism Novice: The German Case

The provision of universal service in basic telephony was a primary telecommunications policy goal in Germany following World War II. The Basic Law required the government to provide universal service directly. With the amendment of Article 87 (f) of the law in 1994, the government is simply obliged to guarantee that universal service is adequately provided.

Section 2 (2) of the Law on Regulating Telecommunications and the Postal Services[12] specifically stated that telecommunications industry regulation was intended to establish and maintain broadly available, modern, and reasonably priced telephone service that was equally available in rural and urban areas, and that users had discrimination-free access to the system. This law was superseded by the *Telekommunikationsgesetz* (TKG) on January 1, 1998.

Identification of Policy Programs

Basically, the TKG has adopted a definition of universal service similar to the one in the American 1996 Act in terms of its broadness. Section 17 (1) of the TKG defines universal service as a minimum set of telecommunications services at an affordable price for the public in respect of which a particular quality has been defined, and to which every service user shall have access, irrespective of place of residence or work.

The federal government has been empowered to determine the scope and quality of these universal services by way of statutory instrument. According to Section 17 (2), the government, with the consents of the *Bundesrat* and *Bundestag*, is to determine what constitutes universal service – the minimum level of quality of service that must be provided at a reasonable price. Technical and social developments are to be considered in defining and determining universal services. Specifically, the National Regulatory Agency (NRA) is empowered to determine whether universal service is actually being provided.

The TKG premises that the greater proportion of universal services will be provided by the market without any intervention by the regulatory body. Only if universal service is not being adequately provided, or in cases where it appears likely that such provision will not be ensured will the regulatory body be allowed to step in. Sections 18 and 19 provide for a special procedure in such cases to ensure that universal services will be provided. All licensees, who operate in the applicable licensed market and have at least four percent market share in the relevant market as defined by Section 22 of the Law Against Restraints of Competition, are obliged to contribute to providing universal service. This obligation extends to other companies constituting a single company together with the licensee operating in the market.

Wherever the NRA determines that universal service is not being provided in a particular market, it must first publish this determination in its official gazette. If within one month following publication no market provider declares its intent to satisfy universal service requirements, the NRA may require a licensee holding a market dominant position in the relevant product and geographical market to provide the required service.

Section 20 of the TKG provides that licensees obligated to offer universal services are entitled to receive compensation, insofar as the licensee shows that the long-term additional costs of the universal service exceed the corresponding revenues. Such costs are computed following a complete calendar year in which a deficit in providing the universal service occurs.

Section 21 stipulates that the compensation is financed by the imposition of a universal service fund on the licensees who operate in the relevant market for the licensed telecommunications service in question and achieve a share of at least four percent of the total sales in that market. The share of the levy is determined *pro rata* based upon the licensee's share of sales. Annually, each licensee must report its revenues in that market to the NRA.

Overall, the TKG takes the same basic approach as the American 1996 Act with respect to provisions of universal service. As in the American case, the German government believes that "universal services targets will be achieved in a competitive market" (Organization for Economic Cooperation and Development, OECD, 1999). Because competition is the key driver of improvements in service quality, penetration, and prices, ensuring effective competition in the telecommunications sector will be crucial. The role of federal authority is restricted only to correct inefficient and distortional practices by carriers when ensuring that all citizens have access to certain services of high quality at prices they can afford.

Consistency with Another Policy Goal

One of the greatest challenges for political and industrial segments of German society is to build a highly industrialized society as a decisive step to an information society (*Bundesministerium für Wirtschaft*, BMWi, 1996). The greatest obstacle is the fact that Germany is a developing nation in cyberspace, even if it is on par with the rest of the OECD countries in traditional communication systems.

The Internet was introduced to Germany in the 1980s with the use of e-mail and news and data transfers via UUCP (Unix to Unix Copy Protocol). However, the usage level and rate of telephone and Internet-related services were low. According to the *First Monitoring Report on Universal Service in Telecommunications in the European Union*[13] (Commission, 1998), the proportion of households with telephone service had not yet passed 90 percent by 1997. Only 26.8 percent of German households had one or more PCs compared with 53.5 percent and 38.3 percent in the Netherlands and Britain, respectively. There were only 10.7 Internet hosts and 61.9 webpages per 1,000 population in Germany compared with 65.4 hosts and 283.1 pages in Finland.[14]

An Online-Offline survey[15] estimated 13 percent of the full-time employed population (including school pupils and students) used at least one online service ("Der Spiegel," 1997). In a similar context, the *BT World Communications Report 1998/9* (British Telecommunications, 1998)[16] showed that most of German telecommunications business leaders thought that Germany was far behind the rest of the world in the extent to which it had a competitive, efficient telecommunications market.

One of the reasons Germany's online market has not developed as fast as in the United States is the fact that Germany has no flat-rate structure for local service, so all calls to ISPs are charged per minute. Both Internet provider and user telephone rates are very high. When *Deutsche Telekom* (DT) roughly doubled the phone rates for the inner-city connections most people used to connect to their Internet access providers in 1996, the estimated connection costs to the Internet were 10 times higher in Germany than in North America.[17] The Online-Offline survey confirmed this point by demonstrating that the chief deterrent standing in the way of market penetration was the high cost of using online services ("Der Spiegel," 1997).

To make Germany a leading online nation, the government has positioned the introduction and reinforcement of the market-based regulatory framework as the most important field of action (BMWi, 1996). Once foreign competition has been allowed to enter the lucrative German telecommunications market, it is likely that Internet access and activity will increase. Indeed, the government has worked towards rules that permit more competition and increase the scope for entrepreneurial activity resulting in the passage of the TKG in 1996.

In accordance with the EU timetable, full deregulation was realized on January 1, 1998. Since then, the governmental program has effectively fulfilled its objective in terms of customers' chances to be connected. Currently, DT has changed from a monopolist administration to a market-oriented corporation and is pursuing an aggressive pricing policy in response to inroads the new telecommunications companies have made into DT's market share since 1998.[18]

On April 1, 1999, DT reduced its rates for T-Online and Internet as much as 50 percent. It also unveiled a new pricing package for T-InterConnect (DT's Internet and Intranet platform) to facilitate Internet startup for its corporate customers. Additionally, rates for both long-distance and city calls have been substantially cut ("T-Online Cuts," 1999).

The German approaches toward both universal service and an information society are identically devised to utilize market forces that can be applied to the actual market situation in a cost-effective way.

Relevance of Policy Programs

Broad guidelines for national approaches to universal service issues in Europe have been provided by the EU framework. The German scheme should be evaluated in that respect.

In the past there was no harmonization at the European level of universal service in the telecommunications sector. The absence of competitive forces in most European countries produced mixed results – positive developments such as the full network digitalization in France or the very high levels of service penetration in Scandinavian nations – but also worrying failures in certain countries in terms of poor service, long waiting lists to get a phone, or long delays in repairing faults.

These different levels of development substantially undermine the goal of strengthening economic and social cohesion set out in the Maastricht Treaty, because the existence of a certain minimum set of services is essential for the development of the internal market. Different levels of service obligations in the member states will hamper the take-off of Europe-wide telecommunication services. The concern to achieve balanced development within the different parts of the Community has been a major factor in developing coherent national approaches to universal service.

The maintenance of provision and financing of universal service in a deregulated environment as well as the concept of the service itself are based on the principles of universality, equality, continuity, and affordability. They appear in a number of EU legislative texts, in particular in the Voice Telephony (Council, 1995; Council, 1997a) and the Interconnection Directives (Council, 1997b).[19] Universal service is specifically defined in the Interconnection Directive as "a defined minimum set of services of specified quality which is available to all users independent of their geographical location and, in the light of specific national conditions, at an affordable price." Currently, universal service obligations comprise connection to the public telephone network – at a fixed location – for telephone, fax, and data communications, as well as the provision of operator assistance, emergency and directory inquiry services (including the provision of subscriber directories), public pay-phones, and, where appropriate, special facilities for customers with disabilities or special needs. By including network access, this definition ensures that users access any service provided over today's telecommunication networks, most notably online services including the Internet.

Affordability is so crucial to the extension of telecommunications services to every citizen that it has emerged as a key element in any concept of universal service, but is not yet an explicit requirement of the Voice Telephony Directive. Affordability remains a matter to be determined at a national level because it is closely linked to specific national circumstances and national policy objectives.

The shift from a monopoly environment to competition suggests that rules are needed to determine the extent to which any financial burden associated with providing universal service is shared among market players. Nevertheless, no requirement is imposed on member states to set up schemes for sharing the financial burden of universal service. Each nation is just required to respect Community law and common rules for calculating and financing universal service without discriminating between competitors when it decides to establish some schemes.

Likewise, the EU framework does not require member states to take specific measures on crucial points of universal service policy. It just provides the general principles necessary to unite member states under a single market. Germany properly used the EU suggestions to formulate its own schemes for universal service so that they do not conflict with European efforts to achieve regional cooperation.

As noted above, the German universal service program is strongly based on the belief that "market forces will supply everybody with telecommunications services at an affordable price and that government intervention will only be necessary if it turns out that this is not the case in certain areas" (Kubicek, 1997, p. 408). Even though effective competition is conceived to benefit both residential and business Internet users in terms of choice, quality, and prices, specific measures have not been well refined for the smooth operation of the market mechanism.

For example, the TKG, in contrast to the American 1996 Act, requires carriers with at least four percent of the market to contribute to the universal service fund. This requirement has been criticized as trading "breadth of contribution for simplicity" (Kress, 1997, p. 590). Although the method is relatively simple to calculate, it may well exclude a substantial number of market participants, especially in the early period of growing competition. For this reason, during the passage of the TKG through parliament, the SPD-governed states and the Green Party asked for the inclusion of every network and service provider as contributors to a universal service fund (Kubicek, 1997, p. 408).

Additionally, the *First Monitoring Report* (Commission, 1998) mentioned that Germany did not provide estimates of the net cost of universal service, even if the Interconnection Directives recommended member states identify the cost of universal service obligations and ensure that it could be adequately financed once markets were open to competition. Germany seems to have decided that its national market can be opened to competition without universal service costs being shared with other market players. Its provisions are in place just for the activation of a universal service fund in the event that substantial costs materialize. In this respect, the German approach was called "abstract-general regulation" (OECD, 1999).

These analyses lead to the conclusion that the German telecommunications program of universal service has external relevancy, but lacks internal relevancy. To put it another way, the program is relevantly set in accordance with the EU's general guideline, but it is too unrefined to take advantage of the strengths of market operation. This might be due to the fact that Germany has little experience in transforming a public monopoly to a market-based system, while the EU framework is imposed as mandatory for the future of Europe.

Unanticipated, Situational Context

To meet its responsibility of ensuring that telecommunications services are available at an affordable price on a nationwide basis, the federal government must guarantee truly competitive opportunities for companies in the telecommunications market – elbow-to-elbow with DT, which has been the dominant force in German telecommunications.

However, bringing about a full competitive environment in the telecommunications sector has not been an easy task, because DT and small, private companies entering the deregulated telecommunications market in Germany conflict with each other in many respects.

The first dispute between them took place in 1997 regarding the interconnection agreements. DT is required to offer to three new competitors direct, unbundled access to "customer access lines," which connect customers to the main switching stations, according to Sections 33 and 35 of the TKG. But DT provided access to the lines by offering only transmission services, thereby forcing new entrants to use the DT's transmission and switching devices. The debate ended with the ruling of the Appellate Administrative Law Court in favor of the three new competitors (Neumann, 1998). After the court decision, DT agreed to withdraw its action and committed itself to submit the requested offer of unbundled access to them.

Despite the court ruling, however, the German government has done very little to ensure competition in the deregulated market. It has protected DT instead of the market (Johnston, 1997, p. 64). New entrants pointed to government approval of DT's plan to reduce prices charged to businesses without passing any savings on to private customers. "In keeping with their desire to protect [DT], Germany's politicians seem[ed] reluctant to change" ("Launching Deutsche," 1996, p. 74) some provisions of the TKG.

But DT complained that the TKG imposed undue burdens on the company, which prevented it from adapting to the needs of the market. Above all, the fees allowing DT to charge competitors for access to its network do not sufficiently reimburse DT for its investments in the network. As a result, DT is in effect subsidizing its competitors, an untenable situation in the long run that may reduce incentives for firms to invest in infrastructure. In addition, the fact that an entrant provider with only a single switch could become a long-distance carrier and undercut DT's rates has the potential to decrease network efficiency and cause bottlenecks ("Failures of the German," 1997).

As long as the interests of two market forces continue to make friction, it will be difficult to achieve the competitive market the TKG proposed. It also decreases the chances that universal service in Germany will be successful because its program significantly relies on the market mechanism.

Combining Deregulation with Sensible Regulation: The Japanese Case

Prior to the privatization of Nippon Telegraph and Telephone Public Corporation (NTT) in 1985, a universal service principle was only implicitly expressed through Article 1 of the 1953 Public Telecommunications Law. It said the public interest required fair provision of telephone services. Under the law, NTT was obliged to furnish rapid and reliable public telecommunications services to all people, without discrimination, at reasonable charges.

Even after the introduction of the deregulatory policy in 1985, there has still been no legal definition of universal service in Japan. Without any explicit funding mechanisms for universal service, Article 2 of the NTT Law stipulates that NTT shall provide the service throughout the country. Only for telephone service, remote areas are subsidized by urban areas, and uneconomic services such as public telephones and directory assistance are subsidized by profitable ones. With the introduction of both competition in the telecommunications market and new digital information technologies, however, the government has tackled the design of a new, advanced universal service requirement on a nationwide basis in the multimedia age.

Identification of Policy Programs

Universal service became a national issue in October 1994 when the Japanese Study Group for Research into Universal Services and Rates in the Multimedia Age (Universal Service Group) was formed. The group carried out a comprehensive study from various perspectives concerning desirable universal services and tariff systems to facilitate new business deployment and access by all people to advanced information services in the coming multimedia age. It issued an interim report in 1995 and a final report on May 31, 1996.

The interim report mainly discussed the telephone rate structure. Because the metered local system was not suitable for computer communications, local flat-rate service was proposed. NTT, considering the rapid rise of peak usage and the capacity problem, introduced in 1995 a local flat-rate only during the off-peak period (11:00 p.m. to 8:00 a.m.).

The final report has two sections titled "Rates in the Multimedia Age" and "Universal Services in the Multimedia Age" (MPT, 1996). The first section noted that "demand-stimulating rates"[20] were necessary and suitable for the coming multimedia age. The adoption of rate systems that are independent of the data volume and more suited to calls of extended duration such as flat-rate system was also recommended.

The second section explained that multimedia services would be indispensable in people's daily lives in the near future. At the same time, it proposed that the scope of universal service – which was then limited to plain old telephone service – be expanded because there was a fear that social inequities might increase due to differences in accessibility.

Therefore, the section classified universal services into two types: multimedia access service and multimedia service. While multimedia access service covers broadband network service and network facility services, multimedia service includes several kinds of application services such as tele-education and tele-medicine. Internet access service falls between them.

What is worthy of mention here is the allowance of governmental involvement in ensuring multimedia services. The services are proposed as forms of regional information plans by various government ministries. For instance, the Ministry of International Trade and Industry projects "new media communities," while the Ministry of Posts and Telecommunications (MPT) sponsors "teletopias," and the Ministry of Agriculture has its own "greentopia" (West, Dedrick & Kraemer, 1997, p. 87). The role of the ministries is to propose the general design of a particular information plan to local governments. If local governments authorize any model, the central government will give tax reductions and low-interest loans to the local government to construct the projected telecommunications infrastructure in a particular region (Sugaya, 1997, p. 180).

In addition to the scope of advanced universal service, the provision of financial support for that service was the focal point of the final report. The report suggested four options: cross-subsidization, a universal service fund, an access charge, and a voucher system, but advocated the universal service fund system as the most desirable way of assuring access to multimedia services in the future, especially in terms of supporting services in remote areas and for handicapped people. In a word, the universal service fund was to become a major vehicle for maintaining universal service in Japan just as it had become in the United States and Germany.

The report's conclusion was drawn after considering four aspects: (1) neutrality of competition, (2) assurance of transparency, (3) business efficiency, and (4) implementation cost and ease of understanding. The second section specifically mentioned that if market conditions were highly competitive, cross-subsidization could not be selected. Neither would an access charge for long-distance carriers be workable because of bypass. A voucher system, which is direct financial support for low-income people, was not likely to be politically popular in Japan.

As a whole, the provision of universal services in Japan depends upon dynamic competition in the regional markets, which is similar to the cases of the United States and Germany. However, the Japanese approach differs from the others in that it allows both central and local governments to be involved more actively in introducing an advanced multimedia system in each region in addition to establishing a neutral, independent support system.

Consistency with Another Policy Goal

The Japanese government is forging ahead with a "secondary info-communications reform" (secondary reform) as a means of reforming socioeconomic structures for the future (MPT Telecommunications Council, 1994). Since the first info-communications reform (first reform) started in 1985 with the privatization of NTT, there have been a significant number of new entrants and a drastic reduction in the number of non-competitive areas of the telecommunications market (OECD, 1999).

This process has promoted benefits and greater convenience for Internet users in Japan. According to the *White Paper* (MPT, 1998), there was increasing use of the Internet in Japan. By January 1998, the number of access points for Internet dial-up IP connections had reached about 4,600 nationwide (p. 40). In 1997, 6.4 percent of households held subscriptions to Internet access services, which was up 3.1 percentage points from a year earlier (p. 4). In addition, as of January 1998, Japan had the second largest number of "hosts"[21] in the world with about 1.17 million or 3.9 percent of the total (p. 40).

But Japan had become a lumbering giant with slow economic growth by the mid-1990s. The telecommunications sector faces an entirely new situation driven by the globalization of the industry, the diversification and sophistication of consumer needs, and the introduction of multimedia services supported by technical innovation. These are growing increasingly important and are major tasks for the future of Japan. These factors, together with the success of the first reform, provided an impetus for the initiation of the secondary reform.

To imbue the telecommunications sector with a new dynamism, the Japanese government implemented a "three-in-one" policy: promotion of deregulation, promotion of network connection, and reorganization of NTT (MPT Telecommunications Council, 1997; OECD, 1999). First, deregulation had already provided Japan's socioeconomic system with a number of benefits such as reduced telecommunications charges and the diversification of services. This resulted from increased competition among businesses brought about by the active entry of a large number of new companies into the sector. The Japanese government anticipates further deregulation will establish a more competitive environment, re-injecting new life into the telecommunications market.

Second, given the fact that other business operators have no choice but to use NTT's regional network and that user convenience should be enhanced, it is necessary to ensure that charges for network use are reduced and that users are provided with as wide a range of services as possible. The advance of telecommunications facilities that handle a large proportion of subscriber lines (designated telecommunications equipment) is under examination for both stimulating competition and enhancing user convenience through promotion of efficient interconnection.

Finally, it is widely believed that monopoly carriers likely lack incentive to cut cost and improve services and that monopoly is detrimental to fair and effective competition. Therefore, NTT is eventually to be separated into a pure holding company with eastern and western regional telecommunications companies and a long-distance company (MPT Telecommunications Council, 1996; OECD, 1999). The regional communications companies will be special corporations responsible for the management of regional telecommunications services, and will be obliged to provide universal service to all parts of the country.[22]

In the final analysis, the three policy programs depend on the market mechanism for obtaining the goal of the secondary reform. But the government does recognize that there are problems that cannot be resolved by this approach alone. It is particularly concerned that inequality will arise between regions and between individuals in their ability to participate in the new information system if the whole thing is left to market forces. Therefore, the Japanese government has taken the lead in establishing the second deregulatory policy in the telecommunications sector. The *Vision 21 for Info-Communications* specifically stipulated the government play the roles of presenter of a clear vision, developer of information infrastructure, promoter of dynamic competition, and ensurer of social fairness (MPT Telecommunications Council, 1997).

In sum, even though the success of the secondary reform relies on the creation of a more active market, the government will not necessarily have a smaller role in the process. Likewise, both the advanced universal service and the secondary reform have been identically projected in terms of government leadership and guidance.

Relevance of Policy Programs

The secondary reform goal also constitutes a higher principle guiding the advanced universal service policy. Indeed, securing universal service was referred to as one of the most important issues in the *Vision 21 for Info-Communications* (MPT Telecommunications Council, 1997).

The report specifically noted that despite the fact that information technology could potentially provide everyone access to information, information disparity had emerged between different geographical areas because the market gives priority to metropolitan areas in establishing the new infrastructure. The information gap also occurs between the information-literate and information-illiterate like the elderly, the handicapped, and the low-income class. This gap, in turn, increases economic disparity. A MPT study found white-collar workers in Japan who were able to use PCs earned 19 percent more than those who could not ("Keyboard Allergy," 1999).

Universal service, assumed to be a new basic human right in the info-communications society (MPT Telecommunications Council, 1995), is designed to fill these gaps by making it easier to access the

Internet and by improving PC and network literacy, thereby creating an environment conducive to the growth of an advanced information society. At the same time, the guarantee of universal service should be in line with the progress of competitive markets in telecommunications, as seen in the first reform, which promoted consumer benefits and greater convenience for users. Likewise, the success of the advanced universal service program is mutually interdependent with the advance of the secondary reform.

The universal service policy feature that most distinguishes Japan from the United States and Germany is the relatively wide latitude granted to the government in the policy-making and implementation process. According to Matsui (1994), "*laissez-faire* capitalism has never been a part of Japanese ideology" (p. 237). Since the *Meiji* era (1868-1912), government has been viewed as being committed to promoting industry and protecting it against foreign competitors. In stark contrast to the United States where governmental regulation is justified only when there is market failure, regulation is deemed justified in Japan even in the absence of market failure (p. 229).

In a similar manner, the Japanese Supreme Court usually upholds government legislation designed to limit competition under the rationale that excessive competition is unhealthy. For instance, in *Marushin Industries, Inc. v. Japan* (1972), the Court upheld zoning restrictions on commercial projects designed to avoid excessive competition. The restrictions required a distance of 700 meters between retail store buildings. Similarly, in three cases[23] involving public bath (*sento*), the Courts ruled against entrepreneurs who wanted to open baths within close proximity to existing establishments. These cases illustrate the ability of the government to limit competition. They also are decisions that "affect broadcast stations, telephone companies, and home pages no less than the *sento* (which, by the way, served as the Internet of its time – the place where people gathered to exchange information)" (Rosen, 1997, pp. 226-227).

As a matter of policy, the Japanese government believes that the public interest is "better served by a few strong competitors, confident of their position, than a free-for-all" (Rosen, 1997, p. 224). An advanced telecommunications universal service reflects this attitude by permitting the government to take responsibility for the creation of a universal service policy suitable for an information society. Thus, the Japanese government's policy program for universal service is justified by a broader social and legal environment as well as appropriately grounded in the higher principle of the secondary reform.

Unanticipated, Situational Context

Because advanced universal service has not yet been fully implemented in Japan, it is too early to witness any conflict in applying its specific programs to actual market situations, but some problems are likely to arise.

First, because the universal service policy is projected to adapt the Japanese people to a new multimedia era, the scope of the service becomes one of the essential topics in creating a new framework. As described above, multimedia universal service was categorized into multimedia access service and multimedia service in the Universal Service Group's final report (MPT, 1996). While the telephone carrier (including NTT and private carriers) is supposed to provide access service, non-telephone companies are to offer multimedia services. For example, tele-medicine can be provided by hospitals and tele-education by schools.

This indicates that defining multimedia universal service is easy, but it is difficult to define who would be service providers and who would be participants of the universal service fund. According to Sugaya (1997), the easiest way to solve such complicated problems is to exclude multimedia service from advanced universal service. But this is so unattractive that the rather broad concept, as defined by the Universal Service Group, is with ease supported by the majority (p. 184).

Because interactive services like the Internet are to be included as one of the advanced services, it is clear that the role of telephone carriers will be limited in comparison with the one in the age of plain old telephone service. This means that, for example, not only has every community to be connected with telephones and equipped with PCs for access to information resources, but information literacy must be improved in the multimedia age. Businesses indicate that the highest priority in promoting electronic commerce in Japan should be given to improving consumers' skills in operating computers and access to the Internet (27 percent), followed by a reduction of telecommunications charges (20 percent) (MPT, 1998, p. 48).

To increase the information literacy rate, it is important for the public to be educated regarding the use of communication networks. Such education requires public financial support because this process of subsidy is beyond the economic marketplace. This is why an active governmental role has not been disregarded in pressing ahead with promoting competition in the telecommunications sector in Japan. But most financial support is "available for facilities, not [the] software" aspect of universal service (Sugaya, 1997, p. 183).

Indeed, the *White Paper* showed that even though over 90 percent of Japan's state primary and secondary schools were equipped with computers in 1996, other essential "software" elements for improving the use of communications networks remained lacking. For example, only 19.7 percent of teachers in state schools were qualified to teach children computer skills as of May 1997 (MPT, 1998, p. 14).

Second, in achieving the goal of advanced universal service, it is essential to eliminate the bottleneck monopoly situation by restructuring NTT to a regional system. However, this is to take much time and effort because the construction of current regional networks during more than 100 years is subject to various types of preferential treatment such as tax exemption and compulsive liability of subscribers for telegraph-telephone bond by law (MPT Telecommunications Council, 1997).

Finally, it is necessary to clarify the cost of universal service in order to establish the fund. Defining the scope of universal service and designing a system for ensuring its provision are closely related to the cost factor. Without a clear definition of quality, affordable universal service can be an academic concept. By doing so, the government must construct a model that can be used to calculate the cost of universal service.[24] But calculating universal service costs in competitive markets still stands as an issue to be solved in Japan. These three elements could weaken the strength of the Japanese universal service programs, unless the government, together with market participants, addresses them before the implementation of the advanced universal service policy.

Summary and Conclusion

All three countries in essence base their universal service programs on market mechanisms. Universal service has become a means of spreading the benefits of market developments to all users, regardless of their location and socio-economic status.

Specifically, the American 1996 Act unleashes and relies on competitive market forces to lower rates for consumers and to accelerate the deployment of advanced services throughout the nation. This is consistent with the government's stated vision for the NII. Given the increasing commercialization of the Internet and the market consolidation trend retarding competition, however, the American case indicates there is a danger in that the goal of universal service may be lost. The regional Bells and long-distance companies have not cooperated in collecting a universal service fund and have employed anti-competitive measures such as offering heavily discounted Internet access to attract customers away from the local ISPs, thereby decreasing access potential.

In Germany, universal service targets are identified in the TKG. This legislation reflects the government's belief that a competitive market can achieve universal service goals in a similar logic to the American 1996 Act. This approach is adequate in terms of a German endeavor toward an information society and an EU's general aim to cohere different national approaches to universal service. But Germany's little experience of regulating a competitive marketplace betrays unrefined measures. Not only does this make financial support schemes implausible, but it also generates conflicts between DT and new competing entrants, obstructing the creation of a wholly deregulated environment.

The Japanese approach is predicated on establishing a competitive marketplace as well, but it permits a significant degree of governmental involvement in realizing universal service obligations. This propensity does not conflict with a larger political and social climate in which excessive competition unbounded by governmental guidance is not believed to be beneficial to the public. Actually, the government itself has had an important role in pursuing the secondary reform as well as in creating a postwar economic miracle in Japan. Even though this approach has the advantage of redressing market failures particularly shown in the American case, it tends to focus more on the hardware side of universal service such as equipping schools with computers and connecting them to the Internet, than on the software side that includes improving information literacy through education – crucial for the prospect of universal service in the multimedia age.

CHAPTER 6

Free Speech on the Internet After Telecommunications Deregulation

This chapter discusses how the technological potential of the Internet is related to free speech rights and how both elements are involved in forming and maintaining democratic societies. It then analyzes the experiences of the United States, Germany, and Japan regarding freedom of expression issues arising out of the proliferation of the Internet within deregulated markets.

Significance of Free Speech on the Internet

According to Carey (1988), communication has two contrasting definitions: a transmission view of communication and a ritual view of communication. Whereas the transmission view is defined by terms such as imparting, sending, or giving information to others, the ritual view is linked to terms such as sharing, participation, association, or the possession of a common faith (pp. 14-23).

This distinction becomes blurred by the development of new communication systems, which have at least a certain degree of interactivity. According to Rogers (1986), interactivity is a desired quality of communication systems because communication behavior based on it "is expected to be more accurate, more effective, and more satisfying to the participants in a communication process" (p. 5).[1]

What is noteworthy in bringing interactivity to communication systems is a fundamental change of traditional power structures developed for a linear communication environment. The old model of public as receivers of information loses its explanatory power in an interactive condition. By creating symmetry of communication between senders and receivers, interactive communications remodel people as both creators and providers of information.

The most recent, profound development for interactive communication is the Internet. Interactive capabilities of the Internet enable every user to be content producer and consumer, speaker and listener, and publisher and reader. S. Miller (1996) said, "the Internet is fully bidirectional" (p. 324). The interactive, bidirectional nature of the Internet contrasts with traditional mass communications media that is predominantly one-way from producer to large audience with only limited and crude forms of feedback. In effect, the United States district court in *American Civil Liberties Union (ACLU) v. Reno* (1996)[2] grouped the most common ways of communication on the Internet into six categories[3] and recognized that the Internet enabled people to communicate their ideas and opinions almost instantaneously in a variety of ways.

The interactive characteristic of the Internet derives from the fact that, from the beginning, it has been a decentralized network. The Internet was born during the 1960s out of an effort by the United States Department of Defense (DOD) Advanced Research Projects Agency (ARPA) to create a non-centralized computer network. The rationale behind the development of the ARPAnet was the DOD's concern for maintaining the military's ability to communicate in the face of the destruction of one or more key links in the chain of command that might happen during a nuclear engagement. The major design constraint in the face of this threat was to create a communications system that was nonhierarchical and geographically dispersed.

Some technical solutions played a key role in the process. One of the technical problems that stood in the way of constructing wide-area networks was solved by the "packet switching" concept. This entails breaking a message up into several equally sized packets, with each packet containing the address of the recipient. The packets could be sent separately and reassembled at their destination. Another technical problem in letting different kinds of computers talk to one another was solved by the "gateway" concept. This allows packet switching to be accomplished between different networks despite the fact that they use different-sized packets and work at different speeds. The gateway concept pioneers the open systems approach (Giese, 1996, pp. 126-127).

The unique nature of this open architecture promotes a great diversity of ideas and information because it allows Internet users to create their own websites, establish mailing lists, or participate in online discussions without eliminating other sites or pages. Neither is any permission from any central authority required in the process. The Internet's technological capability of posting messages to and from everyone with an electronic address without editorial control or the intervention of elected representatives paves the way for an electronic democracy where people "expand the public sphere and decrease the elite sphere of power and influence" (Hacker, 1996, p. 224). In a similar context, Sunstein (1995) observed that, with the introduction of the Internet, "the problem of access to the media is ... greatly reduced" (p. 1784).

The Internet's potential for individual expression is expected to outshine any traditional public forum's power. By linking people, groups, and governments around the world, as Bates (1996) noted, "the Internet obliterates geography" (p. 50). He specified the advantages of the Internet, compared to earlier media, by saying "on the Internet, distance has no bearing on cost (unlike the telephone). The Internet (unlike broadcasting) not only delivers the world to us, but also delivers us to the world" (p. 50).

The decentralized architecture of the Internet is what makes the medium unique and essential for electronic democracy. But it would be of little use if some public interests were not guaranteed by public policies, because the technical features of the Internet do not by themselves lead to building democratic formations. Above all, Internet users' free speech rights are essential to preserve the environment characterized by an abundance of communication opportunities, because securing such rights will increase opportunities for democratic participation that is a prerequisite for electronic democracy.

Three Experiences of Free Speech on the Internet

No single institution or branch of government acts as an all-powerful body to shape public policy on freedom of expression. Nonetheless, as Pavlik (1996) suggested, "it would be impossible to overstate the importance of the courts in shaping public policy in the arena of communications" (p. 255). This section focuses on laws and court rulings that have set the stage for the development of free speech policies in a climate of deregulation of and increased competition in the telecommunications industry.

Marketplace and the First Amendment: The American Case

Any discussion of freedom of speech in the United States must start with the First Amendment because the notion of free speech is embodied in this amendment to the United States Constitution.[4] The First Amendment is the ultimate protection for communicators in the United States. It has also influenced all of communications industries in the nation, and it will continue to permeate most of the biggest communication and free speech issues of tomorrow.

The First Amendment "rests on the assumption that the widest possible dissemination of information from diverse and antagonistic sources is essential to the welfare of the public" (*Associated Press v. United States*, 1945, p. 20). Regarding the media, Americans have shared the assumption that diversity is a cardinal goal for democracy. As Altschull (1995) observed, "one of the primary assumptions held by American citizens is that democracy thrives in part because of the information disseminated by the news media" (p. 5). A lack of diversity in fact and thought is presumed to leave citizens blinded. Accordingly, "diversity and richness in the media are not ornaments of a democracy but essential elements for its survival" (Bagdikian, 1985, p. 97). Even though the First Amendment presumes that free speech is essential for enhancing diversity, it still needs a specific method to pursue the free speech goal.

Identification of Policy Programs

The United States has placed great trust in the compatibility of market competition with the American dedication to freedom of expression. The faith in the free market guided the Federal Communications Commission's (FCC) comprehensive programs of deregulation throughout the 1980s, particularly with the appointment of Mark Fowler[5] as chairman in 1981. This tendency has been adopted by each of his successors, including the Clinton appointee, Reed Hundt, who often promised to pursue a policy that delivered public benefits by deregulating and relying on private market mechanisms.[6] This regulatory history eventually encouraged Congress to devise and adopt the Telecommunications Act of 1996 (1996 Act).

The marketplace of ideas is a metaphor for an environment in which the government's regulation of communications is limited. The marketplace metaphor remains an integral part of contemporary legal reasoning behind the First Amendment. Recently, it is characterized by the restrictive application of the public forum doctrine to the Internet. Because "the constitutional guarantee of free speech is a guarantee only against abridgement by government, federal or state," (*Hudgens v. NLRB*, 1976, p. 513) the amount of First Amendment protection the Internet receives depends on which category of forum it belongs to. The greatest protection for expression has been bestowed on the public forum.

Since the public forum doctrine was first pronounced in *Hague v. Committee for Industrial Organization* in 1939, it has protected many speech activities by producing both public function and government entanglement tests. While public function analysis applies when a private actor takes on the attributes of a government, entanglement occurs when the public and private actors become inextricably linked.

The public forum doctrine has been severely limited since the early 1970s.[7] As long as the doctrine is interpreted restrictively, the private characteristics of the Internet make it difficult for courts to apply First Amendment standards to this new electronic environment. Such restriction bestows only the minimum amount of protection for expression.

In 1996, the status of the Internet was disputed in *Cyber Promotions, Inc. (Cyber) v. America On-Line, Inc. (AOL)*. At stake in this case was the right of AOL to block junk e-mail being sent from

Cyber to AOL subscribers.[8] This case presented a novel constitutional issue of the Internet users' freedom to send commercial e-mail messages unfettered, uncensored by their service providers.

On November 4, 1996, Judge Charles Weiner issued a Memorandum Opinion and Order that AOL was not a state actor and thus there had been no state action by AOL's activities. It followed that junk e-mail was not protected speech under the First Amendment and private Internet Service Providers (ISPs) had the right to block their delivery to subscribers. The decision was grounded in a 1995 court of appeals' dictum in *Mark v. Borough of Hatboro*, in which three distinct Supreme Court tests[9] were utilized to determine whether there had been state action.

The first test is known as the exclusive public function test. To decide whether AOL performed public functions, the court drew some of the factual findings both about the Internet from *ACLU* (1996) and about ISPs for itself. The court concluded, "no single entity ... administers the Internet" (p. 832) and "AOL was and is a private online company ... not owned in whole or in part by the government" (*Cyber*, 1996, p. 438). Based on the findings, the court held that "AOL is not exercising any of the municipal powers or public services traditionally exercised by the State as did the private company in *Marsh* [*v. Alabama* (1946)]" (p. 442).

It also pointed out that *Amalgamated Food Employees Union Local 590 v. Logan Valley Plaza* (1968) giving picketers a First Amendment right to picket within a shopping mall was overturned by *Lloyd Corp. v. Tanner* (1972) in which the Court ruled that a group of protesters did not have a First Amendment right to distribute handbills in a shopping center, because the shopping center was private property and the protesters had adequate alternative avenues of communication. Therefore, the court in *Cyber* (1996) ruled, "AOL is not a state actor under the exclusive public function test" (p. 444).

The remaining two tests overlap one another and concern government entanglement. The court in *Cyber* (1996) easily rejected the application of this test by mentioning "there has been no government involvement in AOL's business decision to institute or reinstitute a block directed to Internet e-mail sent by Cyber to AOL members or subscribers" (p. 439). This reasoning was reaffirmed by Judge Weiner in denying Cyber's motion for reconsidering his ruling of December 19, 1996.

In a similar matter, CompuServe has won a restraining order against Cyber.[10] The order prohibited Cyber from using CompuServe accounts to send e-mail advertisements. Judge James Graham held that Cyber's continued transmission of unsolicited e-mail to CompuServe subscribers created a viable claim for trespass upon personal property. Likewise, under the metaphor of the free marketplace of ideas, the recent legal trend for Internet speech narrowly interprets the public forum doctrine, classifying online service providers as private actors.

Additionally, under the marketplace metaphor, the First Amendment requires a system of unrestricted markets in speech because only this system lets individuals assumed to be independent and rational make unbiased decisions from competing market alternatives.[11] Truth will win out over falsity or evil speech because of the assumption of human rationality. Then, the best solution for preventing objectionable speech is *more* speech rather than any regulatory control. As a result, even some obscene materials should be tolerated in the marketplace of ideas to avoid the greater evil of inhibiting free expression.

To produce an open and free trade in ideas among speakers, the Court in *Reno v. ACLU* (1997) overturned the Communications Decency Act (CDA), a provision in the 1996 Act that designated it a crime to display or transmit indecent or patently offensive material to minors over online computer networks. Justice John Paul Stevens writing for the near-unanimous Court said that any "interest in protecting children from harmful materials" should be recognized, but this did "not justify an unnecessarily broad suppression of speech addressed to adults" (p. 2346).[12]

This reasoning is congruent with current policy that the best way to pursue truth and justice within the American version of democracy is to have as few restraints on the marketplace of ideas as possible. In addition, it explicitly shows that the technically unique nature of the Internet, expressed as a democratic medium where anyone can be "a town crier with a voice that resonates farther than it could from any soapbox," (p. 2344) makes any attempt to censor speech liable to constitutional scrutiny.

Relevance of Policy Programs

The First Amendment does not define free speech rights with precision, allowing various interpretations. It has been particularly interpreted in two different ways that correspond to its two free speech traditions: one is called marketplace model, and the other is an instrumental or a Madisonian view.[13] Whereas an instrumental view stresses the role of the First Amendment in constructing democratic government, a market approach focuses First Amendment interests in promoting individual autonomy and self-actualization.

Baker (1989) is classified as a leading modern proponent of the marketplace tradition. According to him, the First Amendment should be understood primarily as promoting individualistic values (pp. 47-69). From this perspective, the First Amendment has its significance when individual self-fulfillment is advanced through free expression. It endorses government regulation only if it makes speech markets feasible and protects them against market failures.

The instrumental perspective was formulated and well expressed in the writings of Meiklejohn (1960). His premise was "the principle of the freedom of speech springs from the necessities of the program of self-government" (p. 27). In constructing democratic self-government, wise decisions about public issues require that "all facts and interests relevant ... shall be fully and fairly presented" (p. 26). However, he did not believe that all speech must be protected, because "what is essential is not that everyone shall speak, but that everything worth saying shall be said" (p. 26). Thus, from this viewpoint, only speech that serves democratic self-governance deserves First Amendment protection.

In the United States, the marketplace model has received its preeminent exposition for historical and philosophical reasons. When the Bill of Rights was drafted in the late 1780s, the newly independent American citizens were sensitive to and worried about governmental power in relation to free expression because of their history with the English monarchy and its agents in colonial America. Ironically, the people who fled religious intolerance in Europe were so intolerant of others' expression that the British exercised greater control of their colonies. Moreover, the American Revolution was largely perceived to be a "liberal" revolution, with its emphasis on private enterprise and the freedom to contract (Horwitz, 1991, p. 34).

Accordingly, the instrumental approach to First Amendment interpretation has never commanded a majority of the Supreme Court. Instead, as Fraleigh & Tuman (1997) noted, "the marketplace metaphor was developed in a series of judicial opinions by Justices Oliver Wendell Holmes and Louis Brandeis," (p. 13) two of the greatest figures in the entire history of American law. In historic cases in the early part of the century, they wrote extraordinary opinions, rejecting governmental bans on political speech merely deemed to be dangerous. For example, Justice Holmes argued that "the ultimate good desired is better reached by free trade in ideas, and that the best test of truth is the power of the thought to get itself accepted in the competition of the market" (*Abrams v. United States*, 1919, p. 630).

As discussed above, this marketplace metaphor has recently evolved with the narrow interpretation of the public forum doctrine and the ensuing classification of online service providers as private actors. It also determines that the government must tolerate expression that many of its citizens may find offensive or dangerous because suppression of the expression is not an appropriate remedy for the discovery of truth. Instead, the preferred response is to speak out against the disfavored idea. This propensity is in the broader boundary of the traditional interpretation of the First Amendment in that it has provided wide latitude for the private sector to guide Internet policies.

Unanticipated, Situational Context

As the Internet cases have set forth, United States courts do not recognize ISPs as a public forum. This has great implications for freedom of expression or communication on the Internet. For example, an ISP can limit unsolicited e-mail from political candidates which it does not support and can stop e-mail on any particular subject of public concern, because its censoring does not involve governmental action.

Indeed, Prodigy, one of the early commercial ISPs, drew nationwide attention in 1990, when it read and censored the e-mail messages of its members. Despite Prodigy's claim that it was curtailing public postings about suicide, sex, and pregnancy, content restrictions were also imposed on messages considered contrary to Prodigy's corporate interests. At that time, some of Prodigy's subscribers protested an anticipated rate raise and posted their complaints on the Prodigy bulletin board. After Prodigy informed subscribers that they could no longer post public messages about Prodigy's fee policy, the subscribers sent e-mail messages to Prodigy advertisers threatening a boycott. Prodigy responded by canceling the protesters' memberships without notice.[14]

Prodigy argued that it was not a common carrier like a public utility that is required to carry all messages, and that there were other electronic forums available to satisfy the free speech needs of its canceled subscribers. However, the controversy brought about a wide concern that the emerging electronic networks would soon be carved up among private providers without any common carrier obligations to free speech. Berman & Rotenberg (1991), among others, maintained that "Prodigy's dispute with its subscribers shows why, to protect First Amendment rights in the electronic age, we need to press Congress to establish the infrastructure for an accessible public forum" (p. 13).

Another example involved the Microsoft Corporation's firing of Gregory Steshenko, a systems-support engineer, for messages he posted on the Internet, "attacking the political policies of the Ukraine and making allusions to Boris Yeltsin's sexual orientation" (Smith, 1994, p. 93). When the company received a significant amount of e-mail protesting Steshenko's firing, it issued a statement saying that the firing was related to job performance and that the corporation "supports First Amendment rights and does not condone political persecution of any type" ("Microsoft Firing," 1993, p. 1).

More recently, AOL, the world's largest ISP, closed most of its Irish Heritage Forum sites, suspending Irish-related services to many Irish and Irish-American customers (Sheridan, 1998). The censorship took place on December 11, 1998 after repeated postings on Ireland message boards that violated AOL's Terms of Service.[15] Irish members considered the Forum closure an outgrowth of pro-British members' pressure on AOL, which was making intensive efforts to market AOL services to potential customers in Britain. The Irish Forum provided cross-border opinion exchanges without British censorship. It has long been a favorite target of anti-Republican, pro-British Internet users intending to disrupt the flow of peace-process information and to limit Irish-American access to the broadest range of Irish views. For these reasons, closure of the Forum was thought to be politically motivated – an attempt to extinguish a growing firestorm of pro-Republican news and comment in the Forum, which exposed abuse of Irish citizens by British-army personnel.

These examples of commercial providers' censorship of Internet speech could have been avoided had United States courts extensively upheld public function and government entanglement tests. These theories most plausibly link the Internet to public forums. However, as case law suggests, there is not much likelihood that the courts will apply these tests to the Internet because the medium is privately built and owned.

Nevertheless, two cases – *Heffron v. International Society for Krishna Consciousness, Inc.* (1981) and *Pruneyard Shopping Center v. Robins* (1980) – could open the door for a more productive reading of the public forum doctrine at the threshold of electronic democracy. In *Heffron* (1981), the Court developed the concept of "limited public forum" by holding that state fairgrounds could fall under this category. The forum was identified by *Perry Education Association v. Perry Local Educators' Association* in 1983. Unlike the traditional public forum and the nonpublic forum, the forum, called a "limited public forum," was defined to include "public property which the State has opened for use by the public as a place for expressive activity" (p. 45).

The limited public forum was clarified in the 1985 ruling in *Cornelius v. NAACP Legal Defense & Educational Fund*, which stated that "a public forum may be created by government designation of a place or channel of communication for use by the public at large for assembly and speech, for use by certain speakers, or for the discussion of certain subjects" (p. 802). Among the characteristics of limited public forum, the most essential is that a government is not required to create this sphere, as long as the state "intentionally open[s] a nontraditional forum for public discourse" (p. 802). In this case, the limited public forum "is bound by the same standards as apply in a traditional public forum" (*Perry*, 1983, p. 46).

Even though the scope of limited public forum was not widely extended in both *Perry* (1983) and *Cornelius* (1985),[16] these cases provide a necessary and useful distinction between limited public forum and nonpublic forum that may help shape the future of the Internet. After these cases, the critical factors in determining the status of the forum become the state's intention of creating the forum and the principal purpose of the property.

In a sense, the Internet has many of the characteristics of the traditional public forum, as discussed above. Taviss (1992) even anticipated that, because of the superiority of the Internet over other means of communications, electronic forums "may one day supersede traditional modes of expression to the extent that alternatives become scarce or impractical" (pp. 782-783). Even though this potential of the Internet has yet to be endorsed by the courts, such an endorsement would allow courts to recognize the Internet as a limited public forum or as a successor to public streets and public parks.

The other possibility for bestowing public forum status to the Internet and for advancing Internet users' First Amendment rights without infringing on the same rights of private operators is to expand the "shopping mall" cases to the Internet. In *Pruneyard* (1980),[17] the Court held that owners of private property could be constitutionally required to open up their property to speech they would rather not have expressed. In its reasoning, this case is in line with *Marsh* (1946) and *Logan Valley* (1968) in which First Amendment rights were applicable to a privately owned entity performing substantial functions of government, rather than with *Lloyd* (1972) and *Hudgens* (1976).

Since the 1980 *Pruneyard* decision, several state courts have wrestled with the issue of "pitted free-speech rights against those of private-property" (Berger, 1991, p. 633) under their respective constitutions.[18] As the court pointed out in *New Jersey Coalition Against War in the Middle East v. J.M.B. Realty Corp.* (1994), the history of shopping-mall cases illustrates that "the Federal Constitution does not prevent private owners from prohibiting free speech leafletting at their shopping centers because the owners' conduct does not amount to state action" (p. 770). Hence, the majority of courts have denied favoring citizens' speech rights over property rights. However, not all states have allowed shopping centers to prohibit free speech on their premises. Under similar circumstances, in fact, mall owners have lost in California, Massachusetts, Washington, Oregon, and New Jersey.

According to the latest court decision, malls that monopolize public space have at least some obligation to permit free speech (*New Jersey Coalition*, 1994). The holding was based on the essential nature of all regional shopping centers such as "the mammoth size of these regional centers, the proliferation of uses, the all-embracing quality of the implied invitation, and the compatibility of free speech with those uses" (p. 780). In addition to these factors, the court also mentioned that "these regional shopping centers are, in all significant respects, the functional equivalent of a downtown business district," which was "the essential [place] for the preservation of the free speech" (p. 780).

The high level of interactivity of the Internet particularly functions as "an assemblage of networks that offer relatively uninhibited expression to its many users" (Schiller, 1994, p. 21). Public use of the Internet outnumbers those who visit and shop at the mall. According to the Internet Domain Survey,[19] the number of host computers connected to the Internet increased from about 210 in 1981 to approximately 125.89 million in July 2001. The Court in *Reno* (1997) already estimated that "about 40 million people used the Internet" around the world in 1996 and the figure was "expected to mushroom to 200 million by 1999" (p. 2334). In sum, the Internet appears to share features of the shopping center in the relevant respects of wide use, key roles for community discussions and activities, and huge size. Therefore, if courts accept this analogy, individual expression on the Internet can be protected under the rationale of *Pruneyard* (1980).

Consistency with Another Policy Goal

The Supreme Court has interpreted the broad mandate of the First Amendment as a general prohibition against content-based restrictions on speech. This does not mean, however, that all content-based regulations are unconstitutional. The most common examples are laws against fraud, defamation, and obscenity.

In particular, the *Reno* (1997) reasoning has created a heated debate over the constitutionality of the use of filtering software to block obscene materials on the Internet. Government interest in the United States has shifted to an application of widespread filtering software and a universal Internet rating system. Soon after the Supreme Court decision in *Reno*, President Clinton convened a summit about the Internet at which proposals for filtering mechanisms and Internet content ratings were discussed.[20] In a meeting with public interest advocates and industry representatives on July 16, 1997, Clinton and Vice President Gore asserted that user control such as installing filtering software programs, as opposed to CDA-style content regulations, was the best way to help parents keep their children from accessing inappropriate material on the Internet.

In accordance with the governmental scheme, some public libraries, such civic groups as Enough is Enough and American Family Association, and computer users have expressed their intentions to use filtering software on their systems, which makes the software products "a hot market" (Krantz, 1997, p. 48). On the other hand, the American Library Association and free speech advocates like ACLU have consistently argued that the installation of blocking software by a public entity like a library on the Internet is unconstitutional under the First Amendment.

Most challenges against software filters come both from the tools' extension beyond the legally obscene sites to include materials that are only indecent or offensive and from the filters' failure to block inappropriate materials for children.[21] Indeed, the filtering software was supposed to block only materials held to be legally obscene under the rules set by *Miller v. California* (1973).[22] However, given the current status of technological development of filtering software, the nature of the Internet to allow new obscene sites to arise on a daily basis, and the difficulty of separating obscenity from others,[23] no filtering software is expected to provide 100 percent protection from online obscenity.

A *Consumer Reports* survey reinforces this point by finding that the even the best filtering software blocked only one-third of obscene sites. Instead, it tends to censor the sites dealing with news stories about violent, adult-oriented subjects ("Protecting Against," 1997, p. B10). In this respect, Meyer (1995) noted that "neither screening of postings nor blocking of access to sexually oriented bulletin boards can be accomplished without also restricting access to speech that is neither legally obscene nor constitutes child pornography" (p. 1981).

Nevertheless, the introduction of the blocking software has its significance in allowing individuals to determine what information is appropriate or inappropriate at the receiving end by permitting the creation of online ratings services without restricting what other Internet users see. It makes access to Internet content a matter of personal choice based on individuals' own tastes or value sets. For this reason, the way of regulating Internet obscenity through the widespread implementation of filtering software is called "user empowerment approach for Internet content."

These user empowerment technologies provide Internet users with more control over Internet content – the approach regarded as the least restrictive alternative for preserving freedom of expression as well as for shielding children from the relevant harms (Berman & Weitzner, 1995, p. 1635; Kennedy, 1996, p. 593). The method also enables the Internet to grow free from unnecessary and ineffective regulatory interference, letting policy-makers assure full respect for the constitutional protection of freedom of expression. In this respect, it does not conflict with the market-centered approach to deal with free speech issues on the Internet.

Regulation for Democratic Discourse: The German Case

In Germany, freedom of expression has been protected by Article 5 of the Basic Law (*Grundgesetz*).[24] According to Kommers (1997), a major feature of Article 5 is its specificity. The Article "protects no fewer than seven basic rights of speech and defines what is to be protected, establishes guidelines for limiting speech, and takes into account modern techniques of communication" (p. 360).

This characteristic makes "German constitutional law relating to freedom of expression [different] in two fundamental ways from American free speech law" (Black, 1997). First, while the First Amendment provides essentially all types of speech with broad protection, the Basic Law does not establish a right to freedom of *speech*; rather, it permits freedom to express one's *opinion*. Therefore, such speech as factual statements that are known to be untrue can be proscribed without implicating the free speech clauses of the Basic Law.

Second, the Basic Law itself mentions two state interests on whom limitations on freedom of expression can be based: protection of youth and right to respect for personal honor. The latter consideration indicates that freedom of expression is viewed by German law as equally important as, not more important than, other protected rights. The right to personal honor enjoys general priority over freedom of expression, when the expression takes the form of "abusive speech."[25] This tendency was strengthened recently when the German legislature passed a law forbidding speech denying the Holocaust.[26]

Identification of Policy Programs

The recent activities of a wide range of German authorities have shown that the German government is determined to regulate Internet content that violates existing speech restrictions on freedom of expression. Two legislative initiatives undertaken to create specific rules for the Internet and to codify existing laws to ensure their applicability to the Internet are the Interstate Media Services Treaty (*Mediendienste-Staatsvertrag*, MdStV) and the Information and Communication Services Act (*Informations- und Kommunikationsdienste-Gesetz*, IuKDG).

These two multimedia laws went into effect on August 1, 1997, after the states developed the MdStV and the federal government the IuKDG.[27] The enactment of both laws is characterized as a compromise or a tentative agreement between the states and the federal government who had fought a bitter feud over who would regulate new online and multimedia services.[28]

The IuKDG[29] clearly postulates that existing German speech limits can be applied to the Internet and are to be enforced whenever possible. This presupposition becomes clearer in later articles of the law that amend the Penal Code[30] and a civil law concerning the dissemination of ublications morally harmful to youth.[31]

In a similar way, Section 7 (1) of the MdStV explicitly intends to apply existing restrictions on free speech to the Internet including laws for the protection of personal honor. Section 8 (1) specifically restates some of the provisions of the law against racist speech and proscribes Internet content that incites war or genocide, glorifies violence, is pornographic (in accordance with the definition of criminal law), or morally corrupts minors. Section 18 (2) of the law also permits individual states to forbid illegal content on the Internet and to order that access to such content be terminated. In sum, the recent multimedia laws in Germany demonstrate the clear determination of the federal and state governments to deal with freedom of expression issues on the Internet by applying existing speech restrictions to the new medium.

Relevance of Policy Programs

The German attempt to find ways to apply its current speech regimes to Internet content can be justified by the historical, political rationale for freedom of expression in Germany and the consistency with European Union's (EU) larger framework of free speech policies. The Basic Law, adopted in 1949, four years after the fall of Hitler's Third Reich, was designed to assure that the abuses of power and suppression of individuals' rights committed under Nazi rule could never occur again. It opens with guarantees of the fundamental right of personal freedom and self-fulfillment along with the inviolability of human dignity.

In particular, the purpose of expression of opinion has been recognized by the German courts as the conveyance of an intellectual effect on one's surroundings and the contribution to the formation of public opinion (Black, 1997). The exclusion of untruth and harmful or abusive speech from constitutional protection also reveals the belief that such speech damages the quality of democratic discourse, because it increases the risk that the public will be misled, or that the public peace could be endangered. This conception of shielding public discourse from counterproductive speech is in contrast with the prevailing marketplace approach to the First Amendment in the United States.

Instead, the German way of justifying free speech rights overlaps a great deal with the instrumental view of the role of the First Amendment. German attitudes toward regulation were typically expressed in a 1981 decision[32] concerning the constitutionality of free markets in broadcasting. By the 1980s, the spectrum scarcity had become a far less plausible basis for regulation in Germany.

But the Federal Constitutional Court emphatically rejected a pure marketplace rationale by saying that the goal of regulation is to ensure the free formation of opinions. To achieve this goal, freedom from government domination is necessary but not sufficient. Thus, government must attempt to ensure that the diversity of existing opinions finds its greatest possible breadth and completeness through broadcasting and that, as a consequence, comprehensive information will be offered to the public. This requirement called for mandatory legislation, even if the scarcity problem no longer existed. Given this uniquely held belief in the role of speech restrictions for democratic discourse formation, it is no surprise to witness the extensive governmental efforts currently underway to regulate the Internet in Germany.

Prior to the enactment of the new multimedia laws and even after that, German prosecutors actively investigated and filed charges against both ISPs and Internet users for transmitting or providing access to illegal content under existing content laws. The first major case arose in the context of child pornography. In 1995, the Office of the District Attorney for Munich[33] launched an investigation into the German subsidiary of CompuServe, based on a suspicion that the company had violated laws forbidding the provision of the conduit over which illegal content such as child pornography was transmitted. Felix Somm, manager of CompuServe Germany GmbH, was accused of allowing CompuServe customers in Germany to access pornography in 1995 and 1996, before CompuServe blocked access to controversial newsgroups.

On May 28, 1998, Somm was convicted on charges of allowing traffic of pornographic materials over the Internet. The Bavarian district court found him guilty of complicity in 13 separate cases of spreading pornography over the Internet, and gave him a two-year suspended sentence. In the opinion, the judge said that the Internet is not a "vast ocean of data," but nothing more than a "data bank." Because the material was stored temporarily on the server of CompuServe, the access provider had the possibility of blocking this material. Then, the judge concluded that the defendant had abused the Internet for economic purposes putting his interests in maximizing profit before child protection and that the verdict should deter other access providers from doing the same.

Not only have Internet access providers been subject to formal investigation by German prosecutors, but individual Internet users have been targeted by the authorities. Shortly after the Munich investigation, the network service T-Online, a subsidiary of Germany's main phone company, *Deutsche Telekom* (DT), reported that it was blocking access to a website maintained by German-born Canadian Ernst Zündel.[34] He had distributed materials contending that the Nazis' genocidal assault on the Jews of Europe did not take place. He was the focus of an investigation by the Public Prosecutor's office in Mannheim in early 1996 for possible violations of German laws. In contrast to the CompuServe case, the prosecution focused on the producer of the material. Zündel was charged with violation of section 131 of the Penal Code prohibiting the depiction of violence.

Additionally, a 23-year-old student in Bonn was charged by the local district attorney in February 1997 with inciting criminal activities by allegedly loading a terrorist handbook onto the Internet via an American server. German courts have also recognized the criminal responsibility of chatroom members for defamatory statements made during online conversations.[35]

Germany is a member of the EU, which has produced a wealth of reports and studies concerning the regulation of harmful and illegal content on the Internet. The first structured reflections in the context of new electronic services were in October 1996 with the publication of the *Green Paper on the Protection of Minors and Human Dignity in*

Audiovisual and Information Services (Commission, 1996b) and the *Communication on Illegal and Harmful Content on the Internet* (Commission, 1996c).[36]

In both documents, the Commission said that there exists no legal vacuum in any online content. Even if the global nature of the Internet makes the national definitions of permissible content so inconsistent that it weakens national law enforcement, the Commission concluded that the authority to prosecute and punish people responsible for illegal Internet content falls under the existing laws of the member states. In addition, the Commission considered such content as child pornography, trafficking of human beings, racism, terrorism, or all forms of fraud to be criminal and called for an agreement among all service providers to remove content that a government decides is illegal in its country. In this context, the EU has shown recognition of the each member state's intent to apply its own speech law to Internet content, and German regulatory effort does not conflict with this broader principle.

Unanticipated, Situational Context

The CompuServe case sparked an outcry from both industry specialists and civil right advocates concerned that the decision could become a barrier to Internet use in Germany. Paulus Neef, chairman of the German Multimedia Association, has issued a statement saying that the conviction of Somm was a "blow for the multimedia industry in Germany" (D'Amico, 1998). Neef pointed out that the court ignored expert witnesses testifying that it was technically impossible for Somm to block the distribution of all illegal content going through CompuServe's Internet access servers.

Indeed, it is questionable whether this decision is in line with Germany's new multimedia law that went into effect in August 1997. The law does not hold ISPs responsible for illegal content if they do not have the necessary technology to block the material. Article 1 of the IuKDG only holds ISPs liable for illegal content in two conditions: (1) when ISPs themselves create the content; or (2) when ISPs have knowledge of illegal content and blocking access to such content is technically feasible and reasonable. They are not liable for any third party content that they provide access to over the Internet.[37] The respective technical way of filtering out the offending material was not available until February 1996 when CompuServe reinstated all but five of the sites and introduced software allowing customers to block unwanted material.

For this reason, even the prosecutor on the CompuServe case appealed in his final statement for the defendant to be exonerated because he found there was little basis to convict the defendant after hearing the testimony of expert witnesses (D'Amico & Busse, 1998). In effect, Somm was cleared on appeal of pornography charges because the judge determined that it had been technically impossible for the defendant to close the illegal material in question in November 1999 (Sayer & Deveaux, 2000). Nonetheless, because the major purpose of the new multimedia law fails to establish legal certainty for online providers, the dispute had had a chilling effect on the growth of Internet service and on new investments in the German multimedia industry.

On the other hand, the German government's intention to regulate Internet content has proven to be ineffective in one of the most controversial legal disputes. Angela Marquardt, a former leader and current member of the Party of Democratic Socialism (successor to the East German Communist Party), had been charged with provoking the direction of a criminal act by offering a link on her homepage[38] to a Dutch online magazine, *Radikal*, that contained an article dealing with plans to sabotage German railways as a means to hinder the rail transportation of radioactive material. Marquardt's homepage also included her short text that opposed violence in political activism and called for a dialogue on militancy.

Her page was shut down at the demand of the Federal State's Attorney's Office, and the entire German server was asked to block access to the Dutch ISP XS4all for maintaining a website for the magazine. Marquardt was acquitted of some charges in June 1997, because a court in Berlin ruled that she created the link to the magazine before it published its terrorist tips. It also admitted it would be difficult to order Internet users with their own websites to check for illegal content located on sites they link to. But she was ordered to pay a fine of 1,000 marks for distributing reports about an ongoing judicial proceeding over the Internet.

This case illustrates two important points about current governmental efforts to regulate the Internet in Germany. First, the regulatory intention is very strong, and authorities are prepared to pursue not only access providers, but also users who offer access to illegal information. Second, the authorities' methods and current laws are inadequate effectively to close off access to illegal content. Despite the fact that German ISPs shut down Marquardt's homepage at the request of German authorities, the page has quickly become available again with the same links via a different service provider. Moreover, access to *Radikal* by Internet users has greatly expanded around the world (including in Germany), since the site was supposedly closed off to German users. Within hours of the order to block access to the magazine, more than 50 mirror sites were created. The *Radikal* homepage[39] is still accessible via over 50 servers on the Internet.

Consistency with Another Policy Goal

Transforming Germany into an information society has been advocated by the federal government as one of the most important goals for the future (BMWi, 1996). In particular, as the Internet has become recognized as an expanding economic sector with good economic growth and employment potential, considerations of industrial policy have taken precedence in the effort to modernize Germany.

In pursuing the policy goal, the government believes that Germany can realize new opportunities only if some obstacles on the way to information society are removed. To boost the competitive strengths of German industry, those engaged in the market are to be provided certain conditions with a reliable basis for innovation and investment. In this context, a key role must be played by the continued deregulation of telecommunications and the creation of uniform national legal arrangements for the provision and use of new information services.[40] That is why the German government has forged ahead with the implementations of the *Telekommunikationsgesetz* in 1996 and new multimedia laws in 1997.

Even if the industrial policy goal is intended to encourage private-sector input, it has been basically designed and initiated by the government itself. For example, the Federal Ministry of Economics (*Bundesministerium für Wirtschaft*, BMWi), started its 20 million marks initiative to promote widespread use of electronic data exchange and electronic sources of information in October 1997. This was particularly designed to enable small and medium-sized enterprises to have access to efficient information management, because only five percent of the enterprises were connected to the Internet.

The Federal Ministry of Education, Science, Research, and Technology (*Bundesministerium für Bildung, Wissenschaft, Forschung, und Technologie*, BMBF) together with DT and other key players in the multimedia industry began a project, *Schulen ans Netz* (Schools on the Network), in 1997, which aimed to connect 10,000 German schools to the Internet by 2000. At the time, there were just 6,000 schools out of 40,000 with connection to the Internet (Commission, 1998). The BMBF is also providing a total of 80 million marks over three years to expand a high-speed network. This will offer the schools low-cost network access via the German research network, *Deutsches Forschungs Netz* (DFN). The DFN is to be available to the whole of the scientific and educational communities at various access speeds as well.[41]

The CompuServe case occurred in a situation where Germany is eager to develop itself into an information society by attracting future investors. As the case created uncertainty for commercial ISPs' liability, most business analysts anticipated that it could be a serious blow to the German endeavors toward building an information society, tarnishing the image of Germany as an economic center (D'Amico, 1998). Nevertheless, the overall progress in achieving the industrial policy goal is unlikely at odds with the German programs to ensure free speech rights on the Internet in that both efforts rely on governmental intervention rather than market forces.

Morally Inspired Regulation: The Japanese Case

Freedom of expression was not an essentially legitimized value of Japanese Constitutional law until after World War II. When the Japanese Constitution went into effect during the Allied Occupation in 1947, the right to express oneself without improper interference was guaranteed to every citizen for the first time in Japanese history. The key constitutional provision on free expression is Article 21.[42]

The Japanese government has recognized that the Internet drastically changes the way in which information is conveyed and public opinion is formed in every field of political, economic, social, and cultural life. In an advanced information environment where individuals have become empowered to create and transmit information, ensuring the right to originate information via the Internet has been referred to both as a basic human right (MPT Telecommunications Bureau, 1996) and as an essential value to the building of the global intellectually creative society of the 21st century (MPT Telecommunications Council, 1995).

Identification of Policy Programs

The government presents three policy programs to realize its free-speech goal: (1) the principle of self-responsibility, (2) application of existing laws against illegal content, and (3) protection of secrecy of communications (MPT Telecommunications Bureau, 1997). First, it gives the greatest importance to the responsibility and moral awareness of network users and service providers. It is individuals who must be fully aware of the potential impact of publicized information and the responsibilities and risks entailed. The government believes that as Internet users become more aware, many of the problems associated with information distribution over the Internet will be resolved. It is, therefore, proposed to seize every available opportunity to promote training programs in businesses and organizations that are involved with the Internet, so that users can become better educated in the intricacies of the information-based society.

The government also expects ISPs to take voluntary measures to control Internet content. Under the Telecommunications Business Law regarding the posting of illegal and harmful content, ISPs can warn senders, erase such content, suspend use of services, or terminate contracts.[43] Because the development of regulations relates to the issue of freedom of expression, however, the government itself considers its own efforts to develop regulations for information to be inappropriate. The step that the government is currently undertaking is to encourage the formulation of self-guidelines by groups of providers.

Second, while the Japanese government supports self-regulation, it also emphasizes that national laws still apply to the Internet and that self-regulation does not insulate Internet users and service providers from prosecution for violations of existing laws. In a word, what is illegal off-line is illegal online in Japan. Because the Internet is a form of communication using public networks and private lines, and because providers are type I or type II carriers, they are, for the most part, regulated by the laws governing the communications system such as the Telecommunications Business Law.

A third policy program, designed to promote the freedom of speech in Japan, is the protection of secrecy of communications on the Internet.[44] The reasoning is that allowing Internet users to transmit information anonymously is helpful in ensuring individuals' freedom of expression. In particular, guaranteeing the secrecy of expression is expected to increase opportunities for social underdogs to avoid sanctions from the majority and to express their views.

Relevance of Policy Programs

To see the relevance of the three policy programs, it is necessary to examine some social and cultural tendencies that seem appropriate to an understanding of the Japanese legal culture. Every Japanese citizen has some recognized duties under natural justice. The tendency is common throughout Asia and dates back at least as far as Confucius. Confucian duties are perceived to be reciprocal. The "reciprocal duty consciousness," mostly given by inferiors in return for adequately benevolent treatment from above, is essential to the operation of the social system (Beer, 1984, p. 102).

In this context, self-realization is achieved by freely fulfilling duties to other persons with a correlative expectation that others will fulfill their duties to oneself. Freedom carries a pejorative implication of selfish willfulness. The Japanese government's broad principle of self-responsibility concerning the use of the Internet can be understood in this culturally specific circumstance.

The reciprocal duty consciousness also affects the modality of dispute resolution. For example, "conciliation by a third party is often preferred to adjudication in a court of law" (Beer, 1990, pp. 43-44). Thus, officially sanctioned mediation of disputes, both public and private, is well established in Japan. Recently, as the development of the Internet gives rise to concerns about the flow of illegal information, a complaints processing system as an *ex post facto* step has been proposed by the Study Group for the Advancement of the Conditions for the Use of Telecommunications.[45]

Additionally, the fact that Japan is a non-individualist, group-centered society is the most timeworn of all truisms about the country (Rosen, 1997, p. 219). In Japan, the locus of identity is identified in very cohesive groups rather than isolated individuals and diverse views largely emanate from the former. Furthermore, the freedom of expression of groups is better protected than that of individuals. This "groupism" culture values individual reticence and, in many contexts, views aggressive assertion of personal opinion as reprehensible. However, individuals, more than groups, especially on the Internet, need free speech rights that encourage, not merely allow, the expression of their concerns. In this respect, protecting secrecy of communications is a proper policy program to let citizens assert their speech rights under law freely.

There have been several legal cases in Japan involving people who violated the criminal law by distributing obscene images on the Internet and were arrested on charges of exhibiting obscene pictures in public. For example, a man in the metropolitan area was arrested in 1996, on charges of providing obscene images on his homepage that he obtained from net news. He allowed Internet users to browse through his homepage and download the images. The person concerned was found guilty on a charge of publicly displaying obscene pictures under Article 175 of the Penal Code and was sentenced to 18 months in prison, suspended for three years by the Tokyo district court. The sentence was later confirmed (MPT Telecommunications Bureau, 1996).

There is strong empirical evidence indicating that the major policy programs for free speech on the Internet have been supported by the Japanese public. According to an Internet survey (1997 MPT Survey) conducted through the homepage of the Ministry of Posts and Telecommunications (MPT) during the period between October 29 and November 12, 1997,[46] 57.6 percent of the total 257 respondents believed that "promotion of enlightening activities to enhance users' self-awareness" was the most effective measure for curbing illegal and harmful information. In contrast, only 19.1 percent of them preferred "legal regulations on the origination of such information."

The survey also showed that the majority of Japanese Internet users viewed privacy as the most important value. Almost 69 percent of the respondents said that countermeasures of some kind were needed to deal with "information violating the privacy of others." The second most frequently cited statement concerned "fraud on the Internet" (66.1 percent), followed by "information slandering others" (49.8 percent), "dazzling and false advertisement" (47.5 percent), "infringement of copyright" (46.7 percent), "description of excessive violence" (22.2 percent), and "indecent information" (17.1 percent). Likewise, these statistics have revealed some general agreement about the adequacy of such free speech policy programs as promotion of self-responsibility and protection of secrecy of communications.

Unanticipated, Situational Context

In tandem with the government's principle of self-responsibility, some groups have already started to make their own guidelines. Above all, Telecom Service Association (TSA), the organization of type II telecommunication carriers, formed the Study Group of the Illegal or Harmful Content on the Internet in 1996 and studied the guidelines for ISPs. TSA has largely attended to the following points: promotion of the free distribution and sharing of information by respecting the freedom of expression of message senders; protection of personal data by ensuring secrecy of communications; the rights and interests of the message receivers infringed by illegal or harmful content; and balance between freedom of expression and secrecy of communications with the prevention of the flow of illegal or harmful content. Based on the study results gained, TSA summarized the basic ideas and published the draft called *Operation Manual: Guidelines on the Semi-Open Communications Services for Internet Service Providers*.[47]

Concerning the other policy programs, however, some factors raise questions about their effectiveness. Above all, some people maintain that the virtual society created on the Internet is a special space somehow separate from real society and that the rules of real society do not apply to it. Various problems have been identified with respect to the application of specific existing laws.

Among others, there is an argument as to whether "indecent information" falls under the category of "obscene writings, pictures, or other obscene objects" specified in Article 175 of the Penal Code.[48] As Beer (1990) observed, "Japanese society is rather tolerant of erotica in print, in pictures, and in other media" (p. 52). The result of the 1997 MPT Survey reinforced this point by showing that only 17.1 percent of the pollees replied that countermeasures of some kind were required for "indecent information" among the many illegal and harmful contents on the Internet.

It is interesting to note the results of another survey showing that far more regular Internet users approved of "transmitting obscene information" than persons who did not use the medium on a daily basis. The questionnaire survey (1996 MPT Survey), conducted by the MPT for the period from November 8 to November 22, 1996, obtained opinions regarding the Internet from 926 Internet users and 547 members of the general public.[49] It showed that only 17.5 percent of the Internet users recognized "transmitting obscene information" as a major problem, while 66.5 percent of the general respondents said so. Not only does content on the Internet easily spread across national borders, but the Japanese, especially Japanese Internet users, also have accepting attitudes toward pornography. There is a possibility that pornographic material, which is legal in Japan, may make for constitutional problems in countries having stricter definitions of obscenity.

Japan's Penal Code has no provisions for punishing those who have distributed child pornography, because, in Japan, child pornography is not considered child abuse but is merely an indecent sexual act. In other words, the Japanese people do not view child pornography as a serious infringement of children's human rights. But the distribution of the material is an abuse of free speech rights in most all other countries and is, hence, excluded from the fundamental rights guaranteed by constitutions in almost all other countries. This makes it extremely difficult to form a global consensus on the concept of obscenity and penalties attached to it by applying current laws to the Internet in Japan.

There is another concern that secrecy of communications, if abused, might encourage irresponsible behavior, endangering the right to access information freely and thereby hindering the formation of free and accurate public opinion. Indeed, one of the reasons that unreliable or even illegal content is being easily distributed over the Internet is the fact that anonymity hides users' identity.

Therefore, a mechanism must be established to ensure freedom of expression without infringing on speakers' confidentiality on the Internet. Nevertheless, the government fails to provide the specific legal scheme to make the balance between those two different rights. Rather, it just describes the conditions under which the disclosure of confidential information would be permitted.[50] These mere suggestions would not be enough to handle expected conflicts over the scope of secrecy of communications.

Consistency with Another Policy Goal

Japan is an international leader in promoting electronic commerce. According to the *White Paper* (MPT, 1998), total sales of goods and services ordered via the Internet amounted to 28.5 billion yen in 1996, but this figure grew almost threefold to reach 81.8 billion yen in 1997 (p. 46). The Electronic Commerce Promotion Council also estimated that electronic commerce would be worth 15 trillion yen by 2002 ("Japanese Internet," 1999). The figure would be equivalent to one percent of all commercial transactions in Japan.

The use of advanced information technology in economic activities has been primarily pursued by the Ministry of International Trade and Industry (MITI), a rival ministry of the MPT, to improve the efficiency of corporate and industrial activities, even if many governmental agencies have been involved in various aspects over an expansive range of electronic commerce issues.[51] Since 1995, the MITI has implemented 45 projects for consumer-business and inter-business electronic commerce with a total budget of 32 billion yen. It has also established the Committee on the Improvement of the Environment for Electronic Commerce in 1995 to study ways to improve the environment for electronic commerce.

In this endeavor as well, the government indicates that proper measures against morally or socially questionable content are an important issue for the healthy growth of network transactions and the trust of the society in electronic commerce generally (MITI Committee, 1997). As countermeasures to harmful content, both the principle of self-responsibility and the application of existing laws to actual transactions on the Internet have been implemented. For example, the distribution of obscene and defamatory materials is regulated to maintain social order under Articles 175 and 230 of the Penal Code, respectively.

However, across-the-board application of legal regulations is not perceived as a complete answer, especially for ensuring free speech rights while protecting youth from illegal content. Rather, the MITI encourages information suppliers and user groups to exercise self-restraint. Accordingly, the Electronic Network Consortium, an organization of network operators, published two papers in February 1996: the *Guidelines to Ethics* to be observed by network operators and the *Rules & Manners* to be observed by the users.

Likewise, as the Japanese government regards electronic commerce as a key issue to build an "intellectually creative society," along with the right to freedom of expression, it advances similar policy programs toward the promotion of electronic commerce to those for protecting freedom of expression. As a consequence, those multiple goals are expected to progress without any fundamental friction between them.

Summary and Conclusion

In the United States, freedom of expression issues on the Internet have been guided by the fundamental premise currently applied to speech in traditional media: "free speech, uninhibited by government interference, is essential in the advance of knowledge and the discovery of truth" (Horwitz, 1991, p. 32). To guarantee the free expression rights of Internet users, the government must respect the free exchange of ideas on the medium. Even though this free speech policy is adequately rooted in historical, social, and legal traditions, it has negative impacts on Internet users' free speech rights. The American case has noted the proverbial "chilling effect" of commercial providers' censorship on free speech and has emphasized that public forum status should be accorded to the Internet so as to enhance the diversity of discourse in this new electronic environment.

The German case reveals an almost opposite tendency to the American one. Traditionally, Germany proposes some form of content regulation in the media to reserve the essential conditions of democracy, to enhance the quality of rational public discourse, and to ensure that the social structures of discourse remain open. With the implementation of two new multimedia laws in 1997, the German government has demonstrated clear intentions to regulate Internet content. This is justified according to historical, political, and regional considerations. However, it is expected that conflicts between service providers and judicial authorities will increase in the future. Additionally, the regulatory approach to speech on the Internet has proven ineffective because of the open architecture of the Internet.

Instead of intervening in Internet speech directly, the Japanese government has prompted service providers to regulate illegal content according to their own principles. Internet users are also asked to exercise their speech rights in accord with societal moral norms. This approach arises out of culturally specific values and is especially supported by Japanese Internet users. Nonetheless, the method needs refinement because of the widespread of and different scope of illegality of Internet content at a global level. For example, the fact that Japan has no provisions for punishing distributors of child pornography has made for many legal disputes. To cite a specific incident, a diplomat working at the British Embassy in Tokyo was sentenced to three years' imprisonment in 1996 on charges of smuggling many videotapes of child pornography collected in Japan into his own country (MPT Telecommunications Bureau, 1996).

In sum, the approaches to Internet speech in the United States, Germany, and Japan differ in fundamental aspects, and each derives from unique experiences. It is important to note here that the German and Japanese policies are relatively untested because they are designed at a time when the Internet is not widely used by the citizens. Instead, the American experience presents a demonstration of the relationship between the marketplace of ideas and free speech on the Internet. Therefore, this study draws a general conclusion on the consequence of telecommunications deregulation on Internet users' free speech rights mostly by relying on the American case: free speech is not necessarily promoted by granting more latitude to market forces.

CHAPTER 7

Deregulation Within a Political Economy Perspective

This chapter is composed of three sections. The first section draws some conclusions about the relationship between public interest and telecommunications deregulation from the findings in the previous two chapters. The second section situates deregulation policy in a political economy framework and explores its implications in terms of social control. The last section reconsiders the role of the state in realizing public interest in a competitive market atmosphere.

Telecommunications Deregulation and Public Interest

This study has explored a broad spectrum of public interest policy arising from deregulatory movements in telecommunications. Whether promotion of competition can coexist with protection of the public interest remains the subject of continuing debate. Some researchers believe free market forces best serve the public interest because the most efficient way to organize social activity is through a marketplace. Others contend the most effective approach for upholding the public interest is governmental regulation because only the government can check the excesses of private power.

Deregulation in telecommunications raises questions about public interest values because telecommunications is not only crucial for commerce, but has implications for social order and political systems.[1] Deregulation has a revolutionary impact on society, but there is little agreement about the positive relationship between telecommunications deregulation and public goods.

For example, the economist Kahn (1983) assumes that competition better serves users and better protects them from predatory pricing than does regulation. He argued that only the free play of market forces could assure the general good. He urged the end of the public utility concept in telecommunications. On the other hand, Aufderheide (1997), one of the most prominent political economists, said that the issues of the public good in telecommunications development were all large questions that would become ever more urgent in a deregulated environment. She rightly pointed out that the tenet of marketplace proponents that a competitive marketplace was preferable in terms of the public interest was largely untested in telecommunications. For this reason, she advocated that the state play a greater role in promoting public interest.

As both Mody & Tsui (1995) and Mosco (1988) observed, however, much of the literature on the impact of telecommunications deregulation on public interest has been descriptive and prescriptive rather than analytical. This study devoted particular attention to why the three nations in question carried out deregulatory policies and how these affected two public interests – universal service and free speech on the Internet – in consideration of the three general principles of political economy: social totality, power structures, and normative or moral concerns. This context-analytical approach produced the following findings.

The United States has increasingly relied on the forces of market competition to safeguard public interests on the Internet. The assumption has been that the most desirable role the government can play is to remove regulatory barriers to market competition. But the effort to collect a universal service fund is being derailed by the profit motives of commercial interests. Such an economic facet of social totality as the consolidation trend in the Internet marketplace also stifles the formation of a fully competitive market climate that is a precondition for the provision of universal service in the country.

Regarding free speech rights, the American case identified several negative impacts of a deregulatory policy. United States courts adopted unduly narrow definitions of the public forum doctrine in the 1970s. These decisions established a wider level of power structure – the dominance of a marketplace model over an instrumental view of the First Amendment. As this tendency has been applied to the Internet, more latitude has been granted to commercial service providers in dealing with speech activities on the new medium, which constitutes a concrete power relation. These providers acting as commanding social actors tend to restrict speech that runs counter to their interests. This was clearly demonstrated in the experiences of activists on the Prodigy network, layoffs at Microsoft, and closures of Irish websites.

The German case demonstrated relatively quick turnabout in power relations in telecommunications. The main actors in the sector saw governmental intervention as the principal means for achieving public benefits, but they now almost as confidently locate their trust in market mechanisms as a surer path to collective benefit. Specific measures necessary to expand and guarantee universal service by means of market forces have not been adequately designed, making the prospect of securing a universal service fund far less certain.

Free speech issues on the Internet have been approached differently. The German government continues to apply strict, preexisting laws to Internet speech even in an era of worldwide deregulation. In doing so, it shows strong intentions to regulate illegal communications that interfere with democratic discourse formations. This also indicates particular sets of power relations in Internet speech policy in Germany. The approach, however, has generated major disputes over legal liability of network service providers and faces strong opposition from the multimedia industry.

Political economy's concern about the interplay between the economic and cultural dimensions has an important role in the analysis of Japanese policy programs, which has produced interesting results as well. The Japanese government regulates the burgeoning field of the Internet with traditional concepts grounded in an Eastern cultural context. By encouraging the concept of self-responsibility in handling Internet speech, the government has gained wide support from the major participants in the policy arena.

Basically, markets rather than consensus politics are the Western model, but Japan has attempted to blend the two. Deregulation has been considered to be a precondition for the successful transition to a postindustrial society, but deregulatory policy in telecommunications has been initiated and implemented by the government itself in Japan. Although an advanced universal service policy has been mostly intended to utilize market competition, it also has a potential to prevent market failures because the policy allows appropriate governmental intervention in obtaining the universal service goal.

It is too early to draw any definitive conclusions regarding the success or failure of the deregulatory policy from those three experiences. In particular, Germany has relatively little experience with alternative market arrangements of the telecommunications sector. Its deregulatory policy has not made any substantial progress toward the public interest goal, and the advanced universal service policy in Japan is not fully implemented yet. The status of these policy plans makes any evaluation of the operation and results of deregulatory policies tentative, presumptive, and circumstantial.

Nevertheless, a conclusion must be reached if one is to construct reasonable and informed expectations about the likely consequences of deregulatory policy on the public interests. The American experience provides an important insight into what can be expected from telecommunications deregulation. It has had the greatest experience in making use of market forces to pursue public interest goals. Indeed, "most debate over regulation of the Internet has centered around the United States – its place of birth" (Rosen, 1997, p. 214).

Therefore, the overall generalization has been made largely by referring to the American experience. The case for telecommunications deregulation in the United States has rested on the assumption that elimination of regulatory barriers will increase the number of contenders, enhance free competition, and generate market efficiencies. All these ultimately ease the way for realizing public interest values. Do the findings in this study support this logic? In a word, no. This study demonstrated the deregulation argument is empirically doubtful and conceptually flawed from the viewpoint of political economy. The market process has thus far failed to satisfy important and ideal market conditions that are necessary for fulfilling universal service obligations. Neither the promotion of competition nor technological progress is a natural consequence of telecommunications deregulation.

Additionally, the deregulation argument fails to acknowledge the trend toward constriction of a public forum doctrine that could bestow the greatest amount of protection of expression for Internet users. Instead, the tendency has generated restriction of Internet speech by commercial forces seeking to maximize their own private interests. The opportunity for all to participate in the discourse of society through universal service and free speech protections on the Internet cannot be guaranteed in current social totality where telecommunications is fully deregulated.

Deregulation as a Means of Social Control

This section explains why public interests like universal service and free speech are jeopardized in a deregulated marketplace. In doing so, it situates the current deregulatory movements in relation to "social control"[2] and assesses the social and political implications from the perspective of political economy. This section relies on Mosco (1988; 1989), among many political economists, who defined social control as one mode of processing social claims. In particular, he demonstrated some of the mechanisms that modern countries use to process social demands and then showed how these mechanisms are organized into forms of governance.

Mosco (1989) first identified four methods of processing social claims: (1) representation or political power, (2) the market or monetary and exchange power, (3) social control or power derived from socialization, values, norms, etc., and (4) expertise or power based on the possession of information (pp. 88-90). Representation, the first mode, incorporates a wide range of social claims, bringing about "demand overload." It weakens the ability of the society's institutions to govern or to manage the range of demands.

The second method, market power, is based on assigning monetary values to social claims. It tends to restrict claims to those that can be reduced to financial measurement. Market-based societies generally prefer politically and economically powerful groups, which structure the market to process their own social claims.

The third, social control over cultural values and norms, is rooted in the daily experience of people. Because values and norms, formed through the emotional involvement of people, are less easily shaken, social control exhibits a more powerful mechanism for settling claims than the other three methods. The successful inculcation of values, however, requires an institutional apparatus like the family and schools.

Finally, expertise draws on socially sanctioned views of what constitutes correct information, knowledge, and truth. Expert-based decision-making reduces social claims if experts show convincingly that expertise achieves satisfactory results and that only highly trained individuals can claim expert status. However, technological failures like a gas leak incident in Chernobyl can make it difficult to accept expertise as a legitimate mode of settling social claims.

There are different ways to configure these four modes of settling claims. Mosco (1989) organized them in a distinctive way to form two basic conclusions:

1. Representation and expertise are fundamentally different modes. In particular, representation incorporates the widest range of claims; expertise only those that meet strict truth tests.
2. The market and social control are also fundamentally different; the market is an indirect and opaque means of settling claims, social control is direct and explicit. (p. 90)

Mosco (1989) described each of the four methods of settling social claims as concomitant with four basic forms of governance. They are regulation, private competition, expert boards, and corporatism (pp. 90-94). The first form of governance, regulation, offers representativeness within a private market structure. In principle, regulation exercises some managerial control by concentrating the power to define the public interest in a small group of individuals. For example, the *Federal Communications Commission* (FCC) maintaining private markets in telecommunications is responsible for representing the public interest. Regulation fails to efficiently control a political system because it permits too many claims.

Private competition is the second form of governance. It relies on the market to clear social claims. The market is neither self-creating nor self-sustaining. Thus, it privileges experts who make fundamental decisions about the best use of the market mechanism. In a word, the market is a form of governance that usually depends upon public bodies to set the rules about creating and maintaining markets.

Expert boards offer a third means to deal with demand overload. Indeed, expert boards or groups of people whose right to governance is grounded in socially defined expertise reduce the number of social claims. Such bodies proliferating in an advisory capacity throughout governments tighten social control by reducing the legitimacy of social claims conflicting with their own definition of expert-based authority.

Finally, the fourth form of governance increases the steering capacity of the state. Corporatism gives authority to individuals who represent specific components of the economic division of labor. Rather than reducing social claims, this form of governance brings social claims to the fore and gives them voice through economic representation. Figure 2 serves a useful summary of Mosco's discussion of four forms of governance.

<Figure 2> Forms of Governance for Processing Social Claims

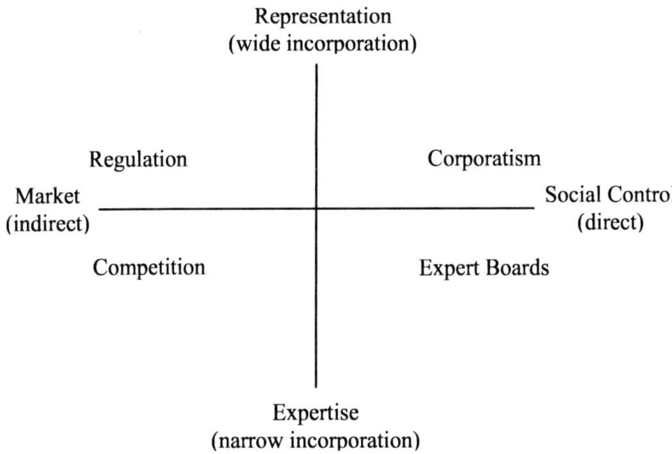

Source: Mosco (1988), p. 110.

This figure broadens our conceptual powers for identifying how power structures in telecommunications have been transformed via a shift in the nature of governance and how control has been exercised by way of deregulation. Government regulation was a typical form of governance in telecommunications, but it permits too many social, political, and economic claims since control over information is of great importance for gaining and maintaining power. Indeed, the German experiences in making universal service and free speech policies have unveiled conflicts between *Deutsche Telekom* and new competing entrants and disputes over the legal liability of service providers in the telecommunications market. The government has transferred claims to the marketplace because of the regulatory tendency to allow many claims and the political incapability to settle them.

Competition as a form of governance departs from the regulatory model by strengthening market control and introducing principles of efficiency into governance. Deregulation is an archetypal method for entrusting the practices of control to the discipline of the market in place of government regulation. It is expected to increase pressure on the market to reduce social claims. In this context, Mosco (1989) described deregulation as "a response to the pressure of mounting social claims on the communications system; it is a means of eliminating social claims" (p. 106). In a word, deregulation can be understood as a transition of primary forms of governance, from government to market.

This transfer of governance is closely related to the political economists' recognition that information or information technology is central to the processes of commodification. Political economy states that most forms of social power come from economic relations. Accordingly, as capitalism develops, material whose initial value is determined by its use, is converted into commodities whose value is decided by what it can bring in exchange. Mosco (1989) especially pointed out that "the history of capitalism is the history of this process, also known as the expansion of the marketplace" (p. 26). This process has expanded into the realm of social production and, at last, has taken over information.

In the wake of the collapse of Soviet Union and its economic bloc, the global balance of power has turned toward the free marketplace. Information, once conceived as a social good that should be shared freely and commonly, has been downgraded into a commodity – an item of economic value. Information is no longer simply a means or resource used in the production of commodities. Instead, it has become a commodity itself that serves only those who can afford to pay a price for it. Likewise, the shift of the form of governance from governmental regulation to market competition as social totality runs parallel to the changing perception of information from a social good into a commodity in political economy.

Indicative of this shift was the conversion of the National Science Foundation's network from public to private hands in the United States beginning October 1994, which is a specific facet of social totality as well. Rapid expansion of the Internet had brought the attention of private enterprise, notably the telecommunications industry, and had awakened the industry to a potentially huge new market. Since then, the

Internet's "expanded use by a much broader range of academics and non-academics, and the development of new, easier to use storage and retrieval capabilities, has changed its character and increased the commercial viability of many of its aspects" (Calabrese & Borchert, 1996, p. 259).

The Internet has grown from a government-funded tool assisting researchers into a competitive business tool. Today, it is almost completely operated by private enterprises. They operate regional network access points, which in turn sell Internet access to local service providers, private enterprises, universities, and government agencies (Baran, 1996, p. 62). Both public organizations like government agencies, research institutes, and universities which rely on government funds and private groups such as corporations and individuals not associated with government funds, pay substantial monthly fees for access to the private companies which serve the Internet (Calabrese & Borchert, 1996, p. 259). In this vein, the infrastructure on which communication via the Internet relies is almost entirely managed by commercial vendors, although "the assumption that the Internet is government-subsidized lingers" (Kahin, 1995, p. 6).

Commercial use of the Internet, which now greatly overshadows its academic and research uses, is another aspect of the social whole. The commercial domain, '.com,' replaced the educational one, '.edu,' as the single largest domain in 1994. As Golding (1996) noted, "the most significant change in the character of the WWW [World Wide Web] has been the irresistible rise of commercial sites" (p. 73). Indeed, the Internet itself has become "a commercial more medium," which is used to communicate marketplace information (Pridgen, 1997, p. 239). Three commodities that corporations are trying to market by participating in the Internet business can be classified as follows: their own goods; access to the network; and advertising[3] (Golding, 1996, pp. 75-76).

In short, the current Internet environment is shaped in favor of private corporations' marketing craving, which is similar to the watershed deregulatory trend in broadcasting during the 1980s, and characterized by the removal of government funding and the creation of new private business opportunities. It is inevitable that commercialization and privatization of the Internet will be the dominant trend in the future.

In replacing regulation with private markets, deregulation puts pressure on the markets to realize the public interest values that regulation sought to achieve and to offset the anticipated failure of the market to meet non-market goals. However, the new arrangement is not likely to succeed from the viewpoint of political economy because control is present in the private marketplace just as it was in the regulated environment.

This control is chiefly exercised through mergers and acquisitions that have increased concentration of ownership in the core of the telecommunications industry. As political economy illustrates, the results of the market logic have been the facilitation of corporate concentration and conglomeration, the development of networks responsive to business user needs, and the commodification of information. These elements tend to monopolize the Internet and hinder the development of competition, endangering the concept of universal service and restricting free expression in a new electronic environment. In this respect, McChesney (1996b) mentioned, "once the needs of corporations are given primacy, the public interest will invariably be pushed to the margins" (p. 103). This tendency has been explicit in the American case.

In a more radical vein, Jansen (1988) proposed a new concept of "market censorship"[4] which operates under market-dominated information-capitalism. Under this system, "access to control over knowledge [and information] becomes a source of surplus value as well as a potential site of social conflict" (p. 168). The shift to information-capitalism has been promoted by the deregulation movement through "accelerating historical trends toward concentration of ownership within information industries" (p. 170). Although the processes of market censorship are subtler and less visible, they result from the fact a concentrated telecommunication corporation controlling significant market share is little interested in the non-market preferences of citizens. Their primary concern is only to satisfy the demands of people within the boundaries of market competition.

Actually, from the view of political economy, the market has proven itself to be a highly flawed regulatory mechanism throughout the history of communication regulation. The historical record of previous media indicates a lot of the same kinds of bright promises about the democratic opportunities inherent in the new medium. For example, "radio was destined to be a 'classroom of the air.' Television in its early days was championed for its potential to educate and illuminate. Cable ... was championed as a boon for health care ... and community interaction" (Shenk, 1995, p. 22). Even though each new medium promised to enhance the democratic ideals, we have ended up being disappointed in many respects as the commercial media have set the rules and limits of the game (Tehranian, 1990, p. 236; "Will Commercial," 1994, p. 44).

This political economy analysis leads to the thesis that market-based deregulation does not mean that control will no longer be an active player. Social control under a deregulatory policy is exercised by the marketplace rather than by the government itself, in terms of Mosco's configuration. Social control becomes an end and the market comes to be a means. For this reason, Mosco (1988) said "deregulation is additionally a political instrument, one that 'unleashes' new instruments of social control" (p. 120). Thus, this section concludes that the deregulatory trend in telecommunications does not mean the abandonment of any kind of control. Rather, it can be understood as a roundabout way of control through market mechanisms.

A New Role for the State in the Future of the Internet

This study started with the observation that the Internet must be bestowed full free speech protections and be easily accessible in societies to augment public discourse and democratic will formation if electronic democracy is a desirable outcome. However, the documents studied in terms of political economy suggest deregulated economic markets are neither a sufficient nor a necessary condition for guaranteeing universal service and free speech on the Internet. These public values are too indispensable to be abandoned. If the democratic future is not to end in an empty pledge, we need "human consciousness and will to reshape institutions" (Tehranian, 1990, p. 70). That is the normative or moral domain in political economy. Not only is political economy concerned about social totality mainly composed of power relations, but it pays attention to normative subjects because the theory was originated from the vision for a more equal, democratic society. That also is the main reason why government regulations have to be re-explored in a new electronic environment. For the Internet to serve public interests effectively, the government must have reasonably broad power to regulate the Internet. This section explores such regulatory forms.

In the United States, government regulations are now regarded as detrimental to economic efficiency, technological development, and freedom of speech. There is much history supporting this point of view, and, therefore, it seems there can be little disagreement about the desirability of excluding government intervention in construction of the new communication order. But the history is only half read.

In the recent past, a governmental role was an indispensable element in encouraging and guaranteeing public interests. Government intervention to assure universal service in telephony was mandated worldwide because of the telephone's unique role in allowing people to keep in touch with one another – creating and strengthening the sense of community ties that was necessary to democracy.

As for free speech rights, governmental intervention was practiced especially in German broadcasting policy because of their Nazi experience. The German Constitutional Court ruled earlier that the government was under a constitutional duty to ensure that a broad array of opinion was presented to the viewing public.[5] Using this stream of reasoning, the Court held that private broadcasters might not be allowed to give any single person or group excessive influence over the formation of public opinion.[6] Instead, the state was under a constitutional duty to control the power of advertisers over the content of private programming.[7] These German cases are revealing in that the Constitutional Court has repudiated marketplace thinking and opted in favor of governmental intervention in understanding its own free speech guarantee to require democratic principles.

Furthermore, in Japan where many private companies are engaged in fierce competition over what is widely accepted as a driving force for rapid economic growth, governmental intervention is far from being seen as the enemy of business. Rather, "free competition is often regarded as threatening to the harmony of society" (Matsui, 1994, p. 229).

Even the United States has occasionally emphasized a governmental role in upholding public interest. The most obvious examples are the fairness doctrine and the must-carry rules. The fairness doctrine required broadcasters to devote a reasonable amount of broadcast time to the discussion of controversial issues and to do so fairly, i.e., to afford reasonable opportunity for the presentation of opposing viewpoints. Whether it conflicted with broadcasters' freedom of speech was at issue in *Red Lion Broadcasting Co. v. FCC* (1969). The Court held that the doctrine was consistent with the broadcasters' First Amendment rights by establishing the spectrum scarcity rationale.[8] The reason the Court applied this theory in justifying the fairness doctrine was that the doctrine would contribute to promoting speech, rather than restricting it. Allowing for opposing points of view could add to the diversity of opinion.

Must-carry rules are included in the Cable Television Consumer Protection and Competition Act of 1992 (1992 Cable Act),[9] enacted by Congress overriding a veto by President George Bush on October 5, 1992. They required all cable television systems offering more than twelve channels to reserve up to one-third of their channel menus for the retransmission of local broadcast signals as part of the cable package they provide to subscribers.[10] Right after Congress approved the 1992 Cable Act, numerous cable operators and programmers initiated a constitutional challenge to the must-carry provisions, arguing that the rules unconstitutionally restricted their editorial discretion.

The Supreme Court in *Turner Broadcasting Systems, Inc. v. FCC* (1997) resolved the difficult First Amendment issue arising from a conflict between two competitive media: the cable industry and local broadcast stations. In a sharply divided decision, the Court upheld must-carry rules, revealing the Court's desire to preserve a multiplicity of sources of information, which had been associated with enhanced diversity, by ensuring the viability of local broadcasters at the expense of cable programmers.

Today, we have witnessed a considerable shift in the prevailing notion about the role of government. The era is known as an age of deregulation, propelled by a widespread reinstatement of liberal economic theory. Because the regulating framework has rested on the liberal idea defining truth itself by reference to what emerges through free exchange of ideas, government intervention has been increasingly seen as the natural enemy of free speech, especially in the United States, since the early 1970s.

For example, the idea of frequency scarcity and the concept of the broadcaster as a public trustee have been called into question. The FCC announced in *Syracuse Peace Council v. WTVH* (1987) that it would no longer enforce the fairness doctrine. On review, the district court in

Syracuse Peace Council v. FCC (1989) upheld the order of the FCC without engaging the constitutional issue, treating it simply as an act of administrative discretion.[11] Fiss (1996) described this tendency as "a marked hostility toward the state and a refusal to acknowledge the role the state can play in furthering freedom of speech" (p. 79).

The regulatory components of universal service have also undergone fundamental adjustments in tandem with organizational changes in the telecommunications industry under deregulation. The provision that formerly required publicly owned monopolies to deliver universal service is no longer adequate in a competitive market. With technological development, in addition, advanced telecommunications has become integral to current societies. The traditional scope of universal service provision should be extended because people who receive only plain old telephone service cannot engage in a full degree of interaction in an information age.

Accordingly, not only must policy-makers decide how universal service should be defined and which firms should shoulder the responsibility of providing the service, but they must determine how universal service obligations should be paid for. As discussed above, all three countries in question established new methods for achieving their universal service goals using market mechanisms after deregulation of telecommunications.

However, from the political economy perspective, the prevalence of the marketplace view "elevate[s] the individual's right to self-expression at the cost of promoting equal and effective citizen participation in the debates and deliberations over the common good," (Abramson & Bussiere, 1995, p. 229) because the market allows for little in the way of political deliberation, yields little attention to public issues, and produces dissent-free, mindless opinions resulting in an inert people. Moreover, the fact that free market proponents are most concerned about government censorship, even when it is market censorship that poses the real threat to public interest, needlessly forces governments to assume the attitude of an onlooker.

This political economy analysis recognizes that some form of government regulation is needed to ensure universal service and free speech rights on the Internet. Indeed, both the failure of deregulation to fulfill its public interest requirements such as universal service and free speech (in terms of wholly societal power relations) and the relation of those interests to the future democratic vision of the Internet (in terms of normative or moral dimension) justify governmental intervention in the Internet marketplace.

As described above, the Internet is expected to play a key role in re-energizing the democratic process and rebuilding a sense of community. Guaranteeing universal service on the medium is such an absolute requirement for any degree of equity that service becomes a necessary and basic condition of citizenship in an information-driven society. Also, in a society linked together by the Internet, the capability of every individual to communicate freely is the functional equivalent of the freedom in the world before electronic media to stand on a corner and speak one's mind.

In order to minimize the effect of deregulated market failures and to maximize the democratic prospect of the Internet, a government intervention supported and shaped by public policies is necessary to fulfill the promise of an electronic democracy under a profit-driven Internet environment. Specifically, some revisions in the current drive for deregulation are recommended, even though deregulation is generally perceived as an inevitable, advanced idea. At the same time, the active involvement of government in the Internet marketplace as a strict public service regulator over private Internet activities should be proposed in the future direction of policy-making.

This is not an entirely new project at all. For instance, Sunstein (1992) proposed various regulatory strategies to redeem and further advance the democratic purposes of free speech including free broadcast time for candidates, financial subsidies for public affairs programming, and rights of reply (pp. 288-289). Sunstein (1988) said those measures were necessary because free expression in broadcasting was threatened by "government 'interaction' that allow[ed] the political process to be excessively influenced by disparities in private wealth and private access" (pp. 1577-1578).

Fiss (1991) argued for some form of content regulation by an activist state to ensure that the social structures of discourse remained open. He also proposed that federal courts be charged with the duty to "preserve the essential conditions of democracy" and enhance the quality of rational public discourse by ensuring that the public was presented with those diverse ideas, speakers, and viewpoints that were "systematically ignored and slighted" by commercial markets (Fiss, 1987, p. 788).

Reuben-Cooke (1993) also proposed that government take a stronger public interest, access-based policy approach to regulation, and that it consider factors such as audience receipt of ideas (e.g., the public's right to information), quantity of services (e.g., guaranteed access for content providers), cost (e.g., greater affordability through subsidies), and literacy (e.g., ability to operate hardware and use software intelligently). In addition, the subject of democratic communities in cyberspace and the possible role of government in supporting them have been of great interest to a number of scholars (Abramson, Arterton & Orren, 1988; Arterton, 1987; Barber, 1984; Rheingold, 1994).

Currently, the Japanese government itself has created the Internet Open Salon "*Jiyu-Kattatsu*" as a forum for the free exchange of views concerning trends in Internet technology and to disperse information about the future of the Internet (MPT, 1997). The Ministry of Posts and Telecommunications decided to establish the salon because the latest knowledge and information about the Internet was not being shared equally, even though the medium had become increasingly important. The salon is expected to serve as a common reference for considering the present status, and the future, of Internet development.

These calls for government regulations are premised on a normative perspective that the public will participate in a full and open debate in a society. The regulations themselves can be credited as furthering, rather than limiting, public interest values. As Fiss (1996) mentioned, this way of seeing and believing requires us to "learn to embrace a truth that is full of irony and contradiction ... that [government regulations] can do terrible things to undermine democracy but some wonderful things to enhance it as well" (p. 83). The democratic future of the Internet can be enhanced better under government intervention in and on the Internet than without it. Horwitz's (1989) insight into the role for the state is worthy of attention: "if we accept that the government is representative of the public ..., government's retreat from regulation constitutes a retreat from democratic process" (p. 284).

CHAPTER 8

Conclusion

This concluding chapter provides a brief summary of the preceding chapters and suggests how the initial idea of this study has been distilled into a conclusion. It also discusses implications of this study for other countries and presents possibilities for the application of the study to future research.

Summary

This study has explicated the broadly defined impacts of telecommunications deregulation on two public interests. It focused on Internet policies because, of the many new and potential communications technologies, the Internet captures the greatest attention with specific reference to a future democratic society. Accordingly, understanding which of the Internet's technological features increase the prospects for electronic democracy and which public interests are involved in realizing that democratic vision is of great significance. This research identified both universal service and free speech on the Internet as essential public goods upon which the hope for constructing electronic democracy rests.

This study selected the relevant experiences of the United States, Germany, and Japan as main areas of research concern because they represent different paths of telecommunications evolvement. The chief part of the research has been an examination of how each nation has endeavored to uphold universal service and free speech regarding the Internet in deregulated marketplaces.

This study based its theoretical framework on political economy. In constructing the framework, it clarified political economy in relation to telecommunications policy and underlying economic theories of regulation.

Much of the policy and legal material was obtained from electronic sources. They were thoroughly evaluated in terms of both the configuration formulated from Dye (1998) and policy evaluation questions modified after Fischer (1980). The findings were then synthesized and discussed in a way that re-framed the role of the state in the new communications environment.

The following findings answer the four primary research questions of the study. The first research question asked what the major reasons for the locomotive drive in the fundamental change in telecommunications regulation were. This study found that the United States has increasingly taken deregulatory measures to deal with breakdown of its traditional legal framework caused by technological convergence. In Germany, regional pressures such as the European Union's initiatives for and successful British experience in telecommunications deregulation eroded German preference for public monopoly. In Japan, governmental intention to incorporate telecommunications into a large-scale information industry was a key factor for the deregulatory trend.

The second research question asked how telecommunications deregulation affected universal service on the Internet in the three nations in question. The following summarizes the major findings in accordance with four policy evaluation questions:

> All three nations attempt to obtain their universal service goals by creating competitive marketplaces, although the degree of dependency on the market within each country varies. In particular, the Japanese approach differs from the others in that it does not entirely depend upon market competition. Instead, it allows governmental guidance and intervention in ensuring universal service. But the Japanese program has not produced any specific results because it has not yet been implemented in actual market conditions.
>
> Each of the three universal service goals is consistent with another policy goal in each country's telecommunications. In the United States, the universal service goal is in the context of the vision for the National Information Infrastructure whose success is

predicated on the effective operation of market mechanisms. By the same token, the German universal service goal is consistent with its endeavor toward an information society, because both goals are to be achieved by encouraging unfettered private investment in the marketplace. Lastly, the universal service goal in Japan is coherently devised to drive for a secondary info-communications reform in which deregulation is reasonably combined with government guidance.

The American universal service program is justified by a politic-economic climate that favors retaining the marketplace of ideas. However, it is unlikely to achieve the universal service goal because competition is not necessarily an outcome of deregulation. The German approach is appropriately grounded in a broader European Union principle, but is too unrefined to take advantage of strengths of market operations. The Japanese policy program is based on the higher principle of a secondary info-communications reform. It is also justified by a social and legal atmosphere that recognizes governmental intervention in redressing excessive market powers.

All three nations' approaches betray unanticipated incidents that weaken the strengths of their universal service programs. The American effort has been stymied by commercial forces more interested in mergers than in contributing to a universal service fund. The German method has provoked conflicts between *Deutsche Telekom* and new competing entrants in the telecommunications sector. The Japanese approach is unbalanced in that it gives less importance to improving information literacy such as teaching the use of communications networks.

The third research question addressed the consequences of telecommunications deregulation on Internet users' free speech rights in the three countries. The following summaries also are the major findings according to the policy evaluation questions:

All three countries have different views on how to guarantee free expression on the Internet. The United States adheres to a hands-off policy in Internet speech markets, believing the free operation of markets can provide adequate speech opportunities. Germany takes a top-down approach, emphasizing a governmental role in increasing diverse viewpoints on the Internet. In contrast, the Japanese government takes a bottom-up approach by urging both service providers and Internet users to regulate illegal content according to their own guidelines.

All three nations' policy programs for Internet speech are relevant in terms of their broader boundaries: traditional understanding of the First Amendment in the United States, historical and regional considerations in Germany, and culturally specific values in Japan. In particular, the Japanese approach is widely supported by service providers and Internet users.

There are exceptional circumstances in all three cases that offset the program objectives. In the United States, the restriction of the public forum doctrine has granted more latitude to commercial providers and as a result, free expression has become more limited by market censorship. In Germany, the governmental effort to regulate Internet speech has brought about major disputes over the liability of service providers. In Japan, the legality of obscenity has created international tension because Internet content easily spreads across national borders.

None of the free speech goals conflict with other policy goals in the three countries. The other policy goals – use of filtering software to block Internet obscenity in the United States, industrial policy goal in Germany, and the promotion of electronic commerce in Japan – are equally essential to the development of the Internet.

The last research question asked how to position the new, changing role of the state in relation to the vision for electronic democracy. This study concludes that the public interest in the two areas, universal service and free expression on the Internet, cannot be realized without sacrificing the viability of telecommunications deregulation. In today's climate of deregulation, universal service will continue to require government action, and Internet users' free speech rights are likely to be enhanced if governments intervene in Internet speech marketplaces.

Implications for Other Countries

This study has been limited mostly by the rapid development of Internet technology, the wide range of Internet policies, and different levels of Internet technology and policies in the three countries. For these reasons, analysis of some of the policy evaluation questions could not be completely parallel among the three countries. Nor could each analysis be totally complete. Even though this study has left some issues unexamined, it lays important groundwork in the field of telecommunications policy and offers implications for other nations.

Above all, this research can be extended to derive lessons for developing countries that are in the midst of restructuring their telecommunications systems in one way or another. Indeed, this study began with the assumption that there were lessons either in the potentially negative aspects of deregulation or in the overall positive assessment of liberalized telecommunications in the three more advanced nations in question. Critical insight into the experiences of telecommunications deregulation as it has evolved in those countries can be gained by developing countries.

It is relevant to note here a set of profound questions the new communication era poses for every country:

> How does a society make wise decisions? How do we determine who we are and what we want to become? How do we organize ourselves and our resources to match these perceptions? These questions are particularly relevant to information and communications policy. A fast-changing society needs access to greater knowledge resources to handle the uncertainties and disruptions that change brings. (Dizard, 1985, p. 120)

Such questions are the most challenging and the most interesting aspects of the current formulation of telecommunications policy in developing countries.

Almost all developing countries have taken deregulatory approaches to telecommunications during the past decade or so, encountering many obstacles on a long and rough transitional path from the old government-led policies to modern market-oriented policies. Because the end goal of telecommunications policies in those countries is deregulation, this study might help their policy-makers avoid the pitfalls of the new era of deregulated telecommunications. Their telecommunications communities do not have to make the same trial-and-error mistakes in creating their own deregulatory policies that have been made elsewhere.

Some findings of this study could serve as lessons from which the telecommunications policy-makers in less advanced countries can learn. First, once the interests of private forces are given primacy, the universal service goal has invariably been pushed to the margins. Second, the speech policy grounded in the marketplace of ideas undermines one of the Internet's most appreciated features of being an open forum for the exchange of ideas. In both respects, the Japanese policy combining deregulation with sensible government intervention within the boundary of cultural tradition seems to be a superior approach in obtaining such public interests as universal service and free speech.

Based on these lessons, this study makes the following recommendations as reasonable and cautionary treatments of telecommunications deregulation in developing countries: (1) governments must intervene in universal service provisions by enforcing service funding schemes if it is to avoid dividing the society into information "haves" and "have-nots"; (2) governments must characterize the Internet as a medium for communications of an open and public nature so as to enhance free expression in the Internet speech marketplace.

Even so, these recommendations must be drawn by making relevant analyses of each developing nation's political, economic, social, legal, and technological situations and specific speculations about the similarities with each of the three cases in issue. This study anticipates that more specific and detailed applications will be made in future studies.

Notes

Chapter 1: Introduction

1. This is not surprising to Garnham (1992) because the field of communications "has been overwhelmingly dominated by concern with the message or text and its effects, rather than with the ways in which the institutions of production and distribution structure the wider social relations of communication" (p. 339).
2. The most representative example is the devotion of an entire issue and a section of two leading communication journals, *Journal of Communication*, vol. 46 no. 1 (1996) and *Journalism & Mass Communication Quarterly*, vol. 75 no. 1 (1998), to "The Net" and "The Future of the Internet," respectively.
3. Such terms have their respective definitions. For example, "privatization" is defined as the elimination of a federal role in providing or subsidizing network services typified by the transfer of state-owned assets to private-sector ownership, and "deregulation" as a reduction in government-set rules. Also, while "commercialization" is regarded as permitting commercial users and providers to access and use telecommunications facilities, "liberalization" means the introduction of competition into monopolized markets. For detailed explanations of those concepts except commercialization, see Noam & Kramer (1994), pp. 275-282. This article identified 11 fundamental building blocks of strategies for telecommunications and then categorized liberalization and deregulation as market structure strategies and privatization as ownership strategies. Because of these differences, it would be a mistake to consider those four concepts as alike. Nevertheless, this study uses them interchangeably in many cases, as privatization, deregulation, commercialization, and liberalization processes of telecommunications do relate to one another in key ways.

4. The most dramatic example of this trend is welfare reform in Wisconsin, where private firms are responsible for reducing welfare case loads. In New Jersey and Tennessee, prisons are operated by private businesses (Wykoff, 1997, p. 14).

5. Reflecting the academic interest in electronic democracy, *Media, Culture & Society* dealt with the political and social implications of electronic democracy in its special issue, vol. 18 no. 2 (1996).

6. Grossman (1995) divided democratic political transformations into three stages. Direct democracy was the earliest form of democracy, originating during the 5th century B.C. in the small, self-contained city-states of classical Greece. Every citizen of ancient Athens could take part in making every decision of state. *Agora*, in particular, was the ancient Athenian forum where citizens gathered to talk, gossip, argue, debate political ideas, and find the weak spots in them. The second transformation is representative government. Originating in the United States a little more than two centuries ago, it is a relatively recent idea in the long span of history. Unlike the ancient Greeks, the Americans entrusted the decision-making power of government to a select group of elected officials, chosen to rule on their behalf. The third transformation of democracy is the electronic democracy or republic.

7. The phrase comes from the remark of Newton Minow. As chairman of the Federal Communications Commission, he called television "a vast wasteland" in 1961 because, despite its promise, television had become a severely distorting influence on American life, dropping the level of communication to the lowest common denominator in the marketplace.

8. Babe (1995) distinguished three distinct, yet related strands of political economy: the liberal, the Marxist, and the institutionalist (pp. 72-75). The liberal approach is essentially different and generally hostile to the Marxist approach. While the former is concerned with laws of economics and sees the market as the appropriate venue in determining the allocation and distribution of wealth, the latter starts with the social relations in which the decisions about wealth take place. In this study, political economy exclusively refers to a contemporary communications theory based on a Marxist approach.

Chapter 2: Review of Literature

1. The most common justification for the reasoning is the problem of natural monopoly. It is defined as an industry in which the production technology is such that one producer can supply the entire market more cheaply than two or

more producers. Examples are the public utilities such as telephony, water, gas, and electricity, where there is a requirement for a network of pipes or cables.

2. Public Law No. 15, 42 Stat. 27.

3. 48 Stat. 1064 (codified as amended in scattered sections of 47 U.S.C.).

4. Cross-subsidization means that tariffs for small users are not based on real costs, but are kept affordable by subsidies from such revenue generating operations as international telephony.

5. The division was made in the second Computer Inquiry by the FCC in 1980. The FCC divided telecommunications services into "basic" services that were to be regulated and "enhanced" services that were excluded from regulation (FCC, 1980).

6. They were Ameritech, Bell Atlantic, BellSouth, Nynex, Pacific Telesis, Southwestern Bell Corp. (later to be renamed SBC Communications), and U.S. West.

7. They included long-distance provider AT&T Long Lines, manufacturer Western Electric, its research arm Bell Labs, enhanced service provider American Bell, and its international products and services division AT&T International.

8. Snow (1995) offered, in considerable detail, an evaluation of the costs and benefits of the AT&T divestiture from the perspective of the 11 years since it entered into force.

9. Public Law No. 104-104, 110 Stat. 56 (codified in scattered sections of 47 U.S.C.). The Act is simple in its basic principles. It covers three domains of communications: local and long-distance telecommunications services, broadcasting, and cable and video services. A special title of the Act, referred to as the "Communications Decency Act of 1996," addresses violence and obscenity on networks. The entire content of the 1996 Act is available at <http://www.fcc.gov/Reports/tcom1996.txt>. It is also included in an appendix of a book written by Shaw (1998), along with a condensed version of the primary features of the 1934 Act. An annotated bibliography about the 1996 Act is available in *Federal Communications Law Journal*, vol. 49 no. 3 (1997), pp. 771-783.

10. Even though telecommunications in the United States are subject to the primary jurisdiction of the federal government, potential conflicts with state regulators remain. On the state level, telecommunications enterprises are typically regulated through state public service commissions. Their activities resemble FCC regulation, but wide variety exists between each of the 50 states. The 1996 Act does not make significant alterations to the existing state-level regulatory structure.

11. Noam (1992) provided an excellent historical analysis of the development of telecommunications in Europe.
12. According to Noam (1992), the DBP's 1980 budget, for example, was significantly above that of Nordrhein-Westfalen, the largest German federal state, which encompassed nearly one-third of the West German population. Its revenues were also much higher than those of any German manufacturing firm. Telecommunications accounted for 70 percent of total DBP's revenue, with postal services 26 percent and banking and financial services four percent (pp. 79-81).
13. Under the framework of the Maastricht Treaty, which came into effect in 1993, the member states of the European Community are in the process of creating a broader European Union where goods, services, capital, and people can move freely across national borders and many social and economic policies will be harmonized. The content of the Treaty is available at <http://europa.eu.int/en/record/mt/top.html>.
14. For reviewing the structure and role of the three institutions, see Bauer & Steinfield (1994), pp. 51-53.
15. Directives as indirect forms of legislation are addressed to all EU countries. Directives lay down policy objectives to be realized by member states through national legislation. Since 1988, the European Commission has issued more than 90 directives.
16. For instance, new consortia such as Mannesmann Arcor and o.tel.o. in Germany, Cegetel in France, and Albacom in Italy were formed to offer competitive telephone services. But the consortia had been forbidden in most EU member states from offering public voice telephony – the biggest and most lucrative part of the service, even though this market was opened in Europe.
17. The Basic Law of the Federal Republic of Germany entered into force on May 23, 1949. In content and style, the Basic Law follows a typical pattern of constitutions adopted by other liberal democracies. However, it was called a "basic law" (*Grundgesetz*) because the Parliament Council did not bestow the term "constitution" (*Verfassung*) on a document drafted to govern a part of Germany for a transitional period that would last until national unification. Even after the German unification on October 3, 1990, the decision has been made to retain the Basic Law as an all-German constitution. Currie (1994) has the text of the Basic law as amended to December 1, 1993; Kommers (1997) includes selected provisions of the Basic Law. Additionally, the full text of the law amended by the Unification Treaty of August 31, 1990 and Federal Statute of September 23, 1990 is available at <http://www.jura.uni-sb.de/law/GG/gg0.html>.

18. *Gesetz zur Neustrukturierung des Post- und Fernmeldewesens und der Deutschen Bundespost* (PostStruktG), June 8, 1989, BGB1. I S.1026. (F.R.G.).

19. Hudson (1997) said that the German approach had shown that "opening up markets to competition is possible without privatizing a state-owned enterprise" (p. 125).

20. This has become the Law on Restructuring the Postal Services and Telecommunications (*Gesetz zur Neuordnung des Postwesens und der Telekommunikation: Postneuordnungsgesetz*, PTNeuOG), June 27, 1994, BGB1. I S. 2325.

21. The German government privatized DT in November 1996 when it sold 26 percent of its shares.

22. BGB1. I S. 1120. (F.R.G.). The entire content of the TKG is available in English at <http://www.regtp.de/en/index.html>.

23. NRA is a term used in European Commission directives referring to state telecommunications regulators.

24. Nippon Telegraph and Telephone Public Corporation Law, Law No. 250 of 1952.

25. *Kokusai Denshin Denwa* Company Ltd. Law, Law No. 301 of August 7, 1953. *Kokusai Denshin Denwa* literally means International Telegraph and Telephone.

26. *Koshu Denki Tsushinho* Law, Law No. 97 of 1953.

27. The MPT was established in 1952, with the responsibility for telecommunications policy and regulatory matters, and the improvement and coordination of telecommunications.

28. Law No. 86 of December 25, 1984.

29. The type I businesses include local network services, long-distance, international, and satellite communications, and cellular services.

30. They are general type II carriers and special type II carriers. While special type II are national or international leased line service suppliers, general type II are smaller, dedicated networks.

31. For instance, free entry has been allowed to the type II industry, and no formal restriction is exerted on pricing or operations in this area. In addition, no restriction is imposed on foreign ownership of a type II carrier, while foreign owners are allowed to obtain, in total, one-third of the shares of a type I carrier.

32. *Nippon Denshin Denwa Kabushiki Kaisha* Law, Law No. 85 of December 25, 1984.

33. The TAC has been set up by the MPT to advise the minister on issues regarding the telecommunications market. It mainly deals with the organization of NTT.

34. Especially see the second chapter of his book, pp. 22-45.
35. It is based on a supply-and-demand model of regulation, and its rationale and popularity are attributed to such economists as George Stigler, Richard Posner, and Sam Peltzman.
36. Stigler (1971) particularly pointed out four main mechanisms enabling business groups to be protected from competition: direct subsidy, control over entry, regulations affecting substitutes and complements, and price-fixing (pp. 4-6).
37. Some political economists advocate that the political as a superstructure is automatically shaped by the economic as a base. This Marxist approach is described as "vulgar Marxism." It has been corrected by numerous scholars who argue for the "relative autonomy" of the superstructure *vis a vis* the base (Jessop, 1990). It means that the economic level is conceived as a necessary explanation of the political, social, and cultural dimensions rather than a sufficient account, even though the dynamics of the communication industries have been understood primarily in terms of their economic determination.
38. It is the oldest of the theories of government regulation. This makes it difficult to trace the origin of the theory. Nevertheless, it is generally believed that the welfare economist Pigou (1920) played a prominent role in advancing the idea.
39. An externality, as a key source of the market failure to achieve efficient outcomes, is any effect of a market exchange that is not intended in the transaction. For a detailed account of externality, see Dahlman (1979). According to him, "the increasing complexity of modern technology and society seems to create yet additional unwanted side effects that require classification on a lengthening list of externalities" (p. 141).
40. According to Horwitz (1989), history actually shows the creation of regulatory agencies as the victorious result of the people's struggle with private corporate interests and as the concrete expression of the spirit of democratic reform. He took two historical phases as its examples. They are the "Granger" period referring to the anti-monopoly activism of the agrarian social movements of the late 19th century and the "Progressive" phase in the early 20th century (pp. 24-26).
41. In this regard, McChesney (1987) eventually declared that the notion that government regulates in some disinterested, public interest fashion to repair market failure has crumbled.
42. Research based on this approach is descriptive and tends to predict the effects of regulation on price and output, the onset of regulation, and the pattern of regulation or deregulation (McCormick, 1989, p. 21).

43. This approach, also known as positive political economy or rational expectations approach, arises out of an attempt to depart from mainstream theories strictly based on an economic framework by applying neoclassical economic models to the political domain and has become the leading theory to thinking about regulation. According to Mosco (1996), the greatest momentum for the theory's growth across a range of disciplines and achievement of a significant amount of public attention was the award of the Nobel Prize in economics to its leading scholars such as George Stigler, James Buchanan, Ronald Coase, and Robert Lucas (p. 254). The primary works of public choice theory include Brennan & Buchanan (1985), Buchanan (1986), and Alt & Shepsle (1990). Over the years, public choice theory has been applied to telecommunications policy questions by Derthick & Quirk (1985) and Crandall (1991). Among others, Crandall, analyzing the effects of the breakup of AT&T and the trend toward competition in the United States telecommunications sector that preceded it, found that the divestiture had improved productivity in the industry. Because competition had not completely replaced monopoly, however, the telecommunications sector remained highly regulated. Citing the problems that occurred in regulated competition in other industries such as airlines, trucking, and railroads, he pointed out the inherent dangers of such regulation in telecommunications as well.

44. Normative media theory addresses how the media ought to operate and what is desirable concerning both structure and performance. In extending it to telecommunications, the primary concern of normative theory would be how telecommunications ought to be operated if certain social values are to be obtained.

45. Mueller (1995) noted that the contribution of communication studies to defining revolutionary changes in the technological and economic structure of telecommunications appeared to be marginal (p. 457). Nonetheless, prominent scholars in this area have consistently called for the active role of communication studies in public policy debates. The most representative example is the devotion of an entire section of *Journal of Communication*, vol. 43 no. 3 (1993) to that subject under the theme of "Connecting Communication Scholarship to Public Policy."

46. The word is borrowed from the French, "*telématique*," a neologism coined by Nora & Minc (1980). Telematics refers to the growing interconnection between telecommunications and computerized data processing.

Chapter 3: Methodology

1. For more detailed explanations of these elementary building blocks of critical social research, see Harvey (1990), pp. 19-32.
2. Most policy researchers distinguished program objectives from policy goals. Whereas policies embody the broad attainable goals that are derived from higher level ideals, programs are grounded in the specific operational objectives derived from the general goals. For a relevant account, see Fischer (1980), p. 184; Chambers (1993), p. 101.
3. For more specific explanations of the methods of identifying goals and objectives, see Chambers (1993), pp. 108-111.
4. See at <http://www.fcc.gov>.
5. See at <http://www.ntia.doc.gov>.
6. The address is <http://europa.eu.int>.
7. The address is <http://www.ispo.cec.be>.
8. The addresses are <http://www.bundesregierung.de>, <http://www.bmwi-info2000.de>, <http://www.mpt.go.jp>, and <http://www.miti.go.jp>, respectively.
9. The address is <http://www.iptp.go.jp/index_e.html>.
10. See at <http://www.infodev.org>.
11. See at <http://www.itu.int/index.html>.

Chapter 4: Motives for Telecommunications Deregulation

1. Exploring the origin of the "marketplace of ideas," Altschull (1995) observed that "the origin of the metaphor is unknown, but it was given expression by Greek philosophers (Aristotle once said, "Plato is dear to me, but truth is dearer still") and by Jesus ("Know ye the truth and the truth shall make ye free")" (p. 6). However, with regard to modern democratic theory, Altschull said, "the concept is most dramatically expressed in John Milton's *Areopagitica* essay" (p. 6). Lowenstein & Merrill (1990) presented five historical ideas that have notably advanced and perpetuated the marketplace model in the United States (p. 216).
2. For detailed explanations, see Pool (1983). The basic theme of the book is how to address the issue of which of the three models – print, common carriers, and broadcasting – will dominate public policy, if the boundaries among them are broken in the coming decades.

3. See highlights of inaugural policy speech by Prime Minister Hashimoto Ryutaro delivered to National Diet on January 22, 1996 ("Prime Minister's," 1996).

4. Kalba (1988) documented how a very broad cross-section of Japanese industry took this objective to deregulate its telecommunications sector (pp. 100-105).

5. Ito (1986) took this relationship as a case study of competition. For a relevant account of bureaucratic rivalry in making telecommunications policy in Japan, see West, Dedrick & Kraemer (1997), pp. 86-88.

Chapter 5: Universal Service on the Internet After Telecommunications Deregulation

1. A backbone is a continental network carrier providing connectivity to regional networks.

2. They were defined as followings:

(1) Access: providing a connection for everyone who wants to plug in no matter where they live or work, and having sufficient capacity for meaningful two-way transmission so that people are not excluded because of their location.

(2) Usability: creating an interactive device and interface, with sufficient power and flexibility to be usable by people desiring to either produce or consume information for a wide variety of purposes so that people are not excluded because of equipment inadequacies or personal disability.

(3) Training: providing adequate training and support in a way that is sensitive to people's backgrounds and is integrated into the institutional context of their everyday lives so that people are not excluded because of lack of skills.

(4) Purpose: making sure that the system can be used to accomplish personally and socially meaningful tasks for most of our population, an infrastructure for personal satisfaction and citizenship as well as economic gain, so that people are not excluded because of the system's inability to meet their needs.

(5) Affordability: making sure that the system is generally affordable and cost effective relative to other alternatives so that people are not excluded because of lack of wealth.

3. For more detailed explanations of conceptual changes of universal service, see Kim (1998a), p. 282.

4. The Joint Board is made up of federal and state-appointed representatives and a state-appointed utility consumer advocate.

5. While the first six were originally outlined in the 1996 Act, the last one was newly advanced by the Joint Board. Competitive neutrality suggests "universal service support mechanisms and rules [that] neither unfairly advantage or disadvantage one provider over another, and neither unfairly favor or disfavor one technology over another." For detailed explanations of each principle, see at <http://www.fcc.gov/ccb/universal_service/fcc97157/97157.html>.

6. While line charges, expressed per line, are created to recover fixed network costs, usage charges, expressed per minute, are intended to recover variable network costs.

7. For a relevant account of the history of the NII, see Kahin (1997), pp. 163-166.

8. At the end of the 1980s, the networks of the Internet were all noncommercial and, if not directly subsidized, were indirectly subsidized by virtue of their free use of the cross-country NSFnet backbone. The services to operate the Internet were supplied under contract to the NSF by a service company, Advanced Network & Services (ANS), which was established by MCI, IBM, and Merit Networks. They sold ANS to America Online at a time when the contract with the NSFnet was terminated in 1995. The sale marks a transition of backbone infrastructure of the Internet from federal funding to full private commercial operation (Baran, 1996, p. 62; Kahin, 1995, p. 6).

9. The survey data is the third profile of telephone and consumer penetration released by the National Telecommunications and Information Administration (1999). At the request of Vice President Gore, the administration surveyed trends in telephone subscription as well as ownership and usage of personal computers and modems by contacting the United States Census Bureau to add questions on personal computer/modem ownership and usage in its 1994 Current Population Survey. Based on the data, the first profile was published in "Falling Through the Net: A Survey of the 'Have Nots' in Rural and Urban America" in July 1995. The second profile updated those survey results, using similar data compiled by the Census Bureau in October 1997, and it was published in "Falling Through the Net II: New Data on the Digital Divide" in July 1998. The most recent report, administration's third in the *Falling Through the Net* series, relies on December 1998 Census Bureau data to provide an updated snapshot of the level of access by Americans to telephones, computers, and the Internet.

10. MCI and WorldCom were key competitors in long-distance telephone service. They also were two of the largest competitors in the Internet backbone market. In particular, WorldCom had played an important role in providing service to discount resellers, who had pushed long distance rates down.

11. Peering arrangements refer to deals between the large ISPs to exchange Internet traffic. The agreements call for both sides to carry each other's data traffic over their own networks at no cost. Smaller ISPs send traffic over the Internet without a fee via a variety of public entry points, which are horribly overcrowded.

12. *Gesetz über die Regulierung der Telekommunikation und des Postwesens* (PTRegG), September 14, 1994, BGB1. I S.2325.

13. This report followed from the first systematic survey of telecommunications service indicators in the EU that was published in March 1996 (Commission, 1996d) and responded to the call of the European Parliament for the Commission to report on a regular basis on universal service in the EU. It also examined for the first time the private and public use of the Internet drawing on information provided by NRAs.

14. The *First Monitoring Report* noted that, in the absence of systematic and comparable information on Internet users, the main pointer to the level and rate of development of Internet services can be such indicators as PC penetration, the growth of Internet hosts, and the number of webpages.

15. The survey is the first national survey to deal exclusively with multimedia development in Germany. It, embracing both western and eastern Germany, was conducted through 8,205 face-to-face and written interviews and published jointly by the magazines *Der Spiegel* and *manager magazin*.

16. The report is based on the survey representing the views of 1,031 senior business directors employed by the top multinational companies in 20 countries around the world. The survey was designed to measure (1) global attitudes towards telecommunications competition, (2) global attitudes towards communications technology, and (3) impact of technology on society. Telephone and face-to-face interviews took place from December 1997 to February 1998.

17. In Germany, there have been repeated calls for special "flat rates" for connections to ISPs and other online services. The proposal is largely grounded in the fact that flat rate local service and the exception of ISPs from contributions to the infrastructure by the FCC have been factors for the high penetration of the Internet in the United States. However, the focus of the debate in Germany has been much more on constitutional and regulatory questions rather than those of pricing and costs (Braun & Schaal, 1997).

18. According to the *1998 Half Year Report*, despite intensified competition in the German core market, the DT's net revenue rose by five percent in the first half of 1998. Net income particularly increased by 18 percent during the period (DT, 1998). It is also reported that with 800,000 new customers (42 percent increase over the previous year) and a total of 840 million calls (plus 70 percent), T-Online recorded record growth in 1998 ("Deutsche Telekom," 1999). This means that T-Online has by far the most subscribers and highest growth rates of all ISPs in Europe.
19. Whereas the Interconnection Directive defines the scope of universal service for funding purposes, the Voice Telephony Directive provides a detailed picture of the level of service users should receive. One of the most important elements introduced by the Voice Telephony Directive is the requirement for member states to ensure that universal service is provided at an affordable price.
20. In this concept, rates would be set low from the initial stage of network implementation to promote the diffusion of multimedia services, thereby expanding the market and consequently lowering the charges paid by users.
21. Computers linked to the Internet are called "hosts," and their increase in number serves to indicate how the Internet has spread.
22. For more detailed explanations of the "three-in-one" policy programs, see the MPT website at <http://www.mpt.go.jp/whatsnew/yellowbook/section6-9812.html>.
23. Judgment of January 26, 1955, Supreme Court, Grand Bench, 9 Keishu 89; Judgment of January 20, 1989, Supreme Court, 2nd Petty Bench, 43 Keishu 1; Judgment of March 7, 1989, Supreme Court, 3rd Petty Bench, 1308 Hanreijihou 111.
24. For more detailed account of calculation of universal service costs and its significance, see Asai (1997).

Chapter 6: Free Speech on the Internet After Telecommunications Deregulation

1. For this reason, over-the-air broadcasting, newspapers, and magazines are trying to be interactive by mostly linking to other media such as telephone or online computer networks. For a relevant discussion, see Hudis (1994), p. 41.
2. This is the lower court decision about the constitutionality of the Communications Decency Act (CDA). All three judges agreed that the provisions of the CDA were substantially overbroad because they forced many

Internet users to forgo constitutionally protected speech or risk criminal prosecution. In this context, this case is called cyberspace's most celebrated case.

3. They are (1) one-to-one messaging such as e-mail; (2) one-to-many messaging such as listserv, that is, automatic mailing list services: It allows communications about particular subjects of interest to a group of people; (3) distributed message databases such as USENET newsgroups: Like listservs, newsgroups are open discussions and exchanges on particular subjects. Users, however, need not subscribe to the discussion mailing list in advance, but can instead access the database at any time; (4) real time communication such as Internet Relay Chat: In its simplest forms, it allows two or more to type messages to each other that almost immediately appear on the others' computer screens. It is analogous to a telephone party line, using a computer and keyboard rather than a telephone; (5) real time remote computer utilization such as telnet: It is to access and control remote computers in real time; and (6) remote information retrieval such as file transfer protocol, gopher, and the WWW: the search for and retrieval of information located on remote computers (pp. 834-836).

4. The First Amendment was written chiefly by James Madison, who had consulted the speech clauses in various state constitutions. It mandates that "Congress shall make no law respecting an establishment of religion, or prohibiting the free exercise thereof; or abridging the freedom of speech, or of the press; or the right of the people peacefully to assemble, and to petition the Government for a redress of grievances."

5. Fowler moved vigorously to obliterate as completely as possible 50 years of American regulatory accretion during his six-year tenure from May 18, 1981 to April 17, 1987. Fowler & Brenner (1982) has been evaluated as the most remarkable work that explained the rationale for adopting a marketplace approach to broadcast regulation.

6. The most representative example can be found at <http://www.fcc.gov/Speeches/Hundt/spreh734.txt>. His remarks were consistent with two overriding goals of the Clinton Administration: to increase competition in telecommunications and to ensure that all Americans share in the benefits of the telecommunications revolution.

7. For detailed explanations of this process, see Kim (1999).

8. AOL was the world's largest commercial online service provider that furnished access and use of its e-mail system and its connection to the Internet to its subscribers. Philadelphia-based Cyber was a small electronic advertising company that provided advertising services for companies and individuals wishing to advertise their products and services via e-mail. Its primary business

was to send unsolicited commercial e-mail messages each day to subscribers of online services including millions of AOL subscribers. On January 26, 1996, AOL sent a letter to Cyber, complaining of Cyber's dissemination of unsolicited e-mail to AOL subscribers over the Internet. The following month, AOL gathered and returned Cyber's e-mail messages *en masse* to Cyber's three Internet Service Providers (ISPs). As a consequence of these e-mail bombs, which brought down the computer system, two of the ISPs terminated their contracts with Cyber, and the other ISP refused to make a contract with Cyber. AOL also had blocked five online sites that served as clearinghouses for commercial mailings. Cyber filed suit against AOL in the Eastern District of Pennsylvania in early June 1996, mainly asserting that AOL's actions improperly interfered with Cyber's First Amendment rights to send e-mail advertisements to AOL members through the Internet.

9. First, the court considered whether "the private entity has exercised powers that are traditionally the exclusive prerogative of the state" (quoting *Blum v. Yaretsky*, 1982, pp. 1004-1005). The second test challenged whether "the private entity has acted with the help of or in concert with state officials" (quoting *McKeesport Hospital v. Accreditation Council for Graduate Medical Ed.*, 1994, p. 524). Finally, the court wrestled with whether "the State has so far insinuated itself into a position of interdependence with ... [the acting party] that it must be recognized as a joint participant in the challenged activity" (quoting *Krynicky v. University of Pittsburgh*, 1984, p. 98).

10. CompuServe of Columbus, Ohio, sued Cyber in 1996, arguing that its host computers were bogged down with junk e-mail and that its subscribers were complaining about this fact. See *CompuServe v. Cyber* (1997).

11. In a similar context, Rainey & Rehg (1996) summarized four principal assumptions of the market-oriented speech policy: (1) market forces are ideologically neutral; (2) the marketplace of ideas is open and readily accessible to advocates of diverse issues of public interest; (3) the increase in the number of information sources made possible by new technologies produces an increase in viewpoint diversity; (4) an unregulated market responds most effectively to the public's desire for information (p. 1937).

12. After the decision in *Reno* (1997), civil liberties brought a case to overturn the government's new law, the Child Online Protection Act (COPA). Enacted by Congress and signed into law by President Clinton in December 1998, the COPA made it a federal crime to "knowingly" communicate "for commercial purposes" material considered "harmful to minors." On February 1, 1999, district court Judge Lowell Reed issued a preliminary injunction protecting Internet speakers from prosecution and fines. Judge Reed wrote in his decision

that "the content on the Internet is as diverse as human thought. Perhaps we do the minors of this country harm if First Amendment protections, which they will with age inherit fully, are chipped away in the name of their protection." For more detailed explanations, see at <http://www.eff.org/pub/Legal/Cases/COPA>.

13. For relevant discussions, see Fiss (1986), pp. 1409-1410; Bhagwat (1995), pp. 176-179; Sunstein (1995), pp. 1759-1765.

14. For a relevant discussion, see Berman & Rotenberg (1991); Foerstel (1997), pp. 118-119; Naughton (1992), pp. 409-411.

15. The boards were cleaned up and reopened on December 28, 1998.

16. The issue in *Perry* (1983) was whether teachers' school mailboxes that were not traditionally a forum for public communication could meet the criteria for limited public forum status. The Court found the teacher mail system did not qualify for the greater protection of this category (p. 46). The question whether the regulations restricting the type of solicitation to nonprofit charities violated the First Amendment was at issue in *Cornelius* (1985). The Court accepted access to the federal workplace represented a type of public property. It, however, did not bestow limited public forum status, because there was no evidence government had intended to open that public property for expression.

17. A student organization had been barred from distributing pamphlets and collecting signatures for a petition within the Pruneyard Shopping Center. The students claimed that the owners had violated their speech rights as guaranteed by the California Constitution (p. 77). The Court upheld the students' rights of expression on private property by relying its reasoning on the public's heavy use of the mall ... and the consequently minimal impairment in value represented by the added presence and speech of the students (pp. 82-84).

18. See Berger (1991), pp. 634-636; Steinglass (1996), pp. 1123-1124; *New Jersey Coalition Against War in the Middle East v. J.M.B. Reality Corp.* (1994), pp. 769-770.

19. The survey is regularly conducted by Network Wizards. The data are available at <http://www.nw.com>.

20. For reviewing remarks by the President at this event, see at <http://www.whitehouse.gov>.

21. For reviewing how filtering software works, see Kim (1998b).

22. Since the *Miller* decision, the First Amendment has not protected speech that (1) appeals to the prurient interest by contemporary community standards when taken as a whole, (2) depicts, in a patently offensive way, sexual conduct specifically defined by state law, and (3) lacks serious literary, artistic, political, or scientific value (p. 24).

23. Sexual speech could be so entangled with political, literary, artistic, scientific, or even educational one that it is impossible to separate them clearly. For example, explicit safer-sex educational information would not be obscene, even though it contains some materials depicting sexual activities. Sometimes, important social issues could be presented by using obscene material.

24. It provides the following:

(1) Everyone has the right to freely express and disseminate his opinion in speech, writing, and pictures and to inform himself without hindrance from all generally accessible sources. Freedom of the press and freedom of reporting by means of broadcasts and films are guaranteed. There shall be no censorship.

(2) These rights find their limits in the provisions of general statutes, in statutory provisions for the protection of youth, and in the right to respect for personal honor.

(3) Art and science, research, and teaching shall be free. Freedom of teaching shall not release anyone from his allegiance to the Constitution.

25. According to Black (1997), abusive speech is narrowly defined by the German courts as that which is intended not to address a topic of debate, but to attack a person himself.

26. The law, passed as the form of an amendment to Section 130 of the Penal Code in 1994, provides for imprisonment for up to five years for anyone who publicly denies or portrays as harmless an act of genocide committed under the rule of National Socialism in such a way that is suited to cause a breach of the peace.

27. The IuKDG was originally called the "Multimedia Law," and a preliminary draft was released in July. Since it was felt that the term "multimedia" was too vague, the title was then changed to "Information and Communication Services Act."

28. While the states argued that the Internet and other online services should fall under their authority to regulate the media and broadcasting, the federal government assumed the Internet as under its authority to regulate interpersonal communication and commerce (Braun & Schaal, 1997). With the introduction of the laws, media services have been differentiated from tele-services. The MdStV is only concerned with media services that are offered to an unspecified, general public such as tele-shopping, videotext, and media on demand. The IuKDG regulates information and communication services which are designed for the individual use and which are based on transmission by means of telecommunication. They include tele-banking and tele-medicine.

29. The IuKDG is composed of 12 Articles; in some cases the articles themselves constitute individual laws, while in others they amend preexisting laws.
30. See the IuKDG Article 4 Amendment of the Penal Code.
31. See the IuKDG Article 6 Amendment of the Act on the Dissemination of Publications Morally Harmful to Youth.
32. 57 BVerfGE 295 (1981). The case can be found in Kommers (1997), pp. 408-413.
33. Munich is the only German city with a dedicated team of police officers assigned to monitor Internet content.
34. Based in Canada, Zündel is a self-proclaimed "Holocaust revisionist" and widely believed to be a major international purveyor of anti-Semitic hate literature.
35. *Amtsgericht Rheinbach*, February 12, 1996. 2 Ds 397/95 (Rheinbach Municipal Court). The court held that calling somebody a "slut" online is defamatory, even though the online talk forum is labeled as chauvinistic.
36. While the former takes a horizontal approach initiating a long-term reflection on harmful content across all electronic media, the latter gives policy options for immediate action against Internet content.
37. This aspect of the law was welcomed by privacy and Internet activists group around the world after the earlier uproar over Bavaria's crackdown (Charles, 1997, p. 6).
38. It was accessible at <http://yi.com/home/MarquardtAngela/OLD/index.htm>.
39. The electronic address is <http://www.xs4all.nl/~radikal>.
40. BMWi (1996) indicated, for instance, that the fulfillment of certain conditions in the legal, economic, and technological fields could create up to about 1.5 million additional jobs in Germany by 2010.
41. On the other hand, both the BMWi and the BMBF have built up an initiative, "Forum Info 2000," to stimulate a positive attitude toward the information society on the part of broad sections of the population. For reviewing more governmental projects, see BMWi (1996).
42. "Freedom of assembly and association as well as speech, press, and all other forms of expression are guaranteed. No censorship shall be maintained, not shall the secrecy of any means of communication be violated."
43. See, in particular, Articles 3, 4, 7, and 34.
44. Currently, information on the Internet has been protected under the secrecy of communications specified by Article 4 of the Telecommunications Business Law. The scope includes not only the content of communications but

also associated information such as the names of transmitters. Even though content has been disclosed to the public, the name, address, and location of the sender are considered confidential. Therefore, except for cases when search warrants have been issued from law enforcement authorities, ISPs do not disclose the identity of the sender.

45. The study group submitted to the Ministry of Posts and Telecommunications a report, *The Complaints Processing and Counseling System for Telecommunications Services*, on December 26, 1996. For a brief outline of the report, see at <http://www.mpt.go.jp/pressrelease/english/telecomm/news7-21-2.html>.

46. It is available at <http://www.mpt.go.jp/policyreports/english/telecommunications/rules_reportref_e.html>.

47. The draft is available at <http://www.mpt.go.jp/whatsnew/e-manual.html>.

48. Article 175 provides that:

A person who distributes or sells obscene writings, pictures, or other obscene objects or who publicly displays the same, shall be punished with imprisonment at forced labor for not more than two years or a fine of not more than 5,000 yen or a minor fine. The same applies to a person who possesses the same for the purpose of sale.

49. It is available at <http://www.mpt.go.jp/policyreports/english/group/Internet/contents.html >.

50. They are (1) sufficient precautions must be taken so that confidential information might not be easily disclosed or abused, (2) only a minimum amount of information should be disclosed, and (3) during the process of disclosing the information, sufficient care must be taken to ensure that any irrelevant information or other users' information is not disclosed inadvertently (MPT Telecommunications Bureau, 1997).

51. For instance, matters related to electronic money are handled by the Ministry of Finance; interpretation and application of civil and commerce laws are the concern areas of the Ministry of Justice; and network-related crimes are dealt with by the National Police Agency. To coordinate the opinions from these agencies and ministries and speed up the implementation of electronic commerce, the government set up the Electronic Commerce Subcommittee within the Cabinet in September 1997.

Chapter 7: Deregulation Within a Political Economy Perspective

1. It is interesting to note that the process of telecommunications deregulation is particularly unique in that much of the change in its regulation is a result of technological innovation, which is not found in the deregulation of other industries. In this vein, Horwitz (1989) said, "new technologies in part underlay a change in the policy-oriented *discourse* about regulation" (emphasis in original, p. 221).
2. All societies contain forms of control. All countries, regardless of their political systems, exercise social control over the communications sector. Therefore, as Comstock (1982) pointed out, "there has been no topic more central in the scholarly and empirical examination of the mass media than their role in social control." He also noted that social control is "a topic that has been treated far more often implicitly than explicitly" (p. 205).
3. Advertising on the Internet has taken four basic forms: (1) spamming, the electronic equivalent of junk mail; (2) classified advertising services; (3) commercial advertising sites on the WWW; and (4) so-called cyber malls (Pridgen, 1997, pp. 239-243).
4. Jansen (1988) mentioned that market competition produced market censorship. Private ownership of the media generates private caprice. Those who control the market sphere of producing and distributing information determine what products will be worthy of producing and, thus, which opinions officially gain entry into the marketplace of opinions. For more detailed explanations, see pp. 131-178.
5. 12 BVerfGE 205 (1961).
6. 73 BVerfGE 118 (1986).
7. 83 BVerfGE 238 (1991).
8. Broadcast regulation had been permitted on the grounds that the electromagnetic spectrum was a scarce social resource or a public utility. As a result of this scarcity, station licensees should be public trustees of the publicly owned airwaves. Keeping broadcasting service responsive to public needs required federal government scrutiny.
9. Public Law No. 102-385, 106 Stat. 1460 (codified at 47 U.S.C.).
10. The systems with 12 or fewer channels must program a minimum of three local stations. Public television stations are automatically carried.
11. Congress made several efforts to pass legislation reinstating the fairness doctrine, but they were thwarted by veto threats by Presidents Reagan and Bush (Fiss, 1996, p. 91)

Bibliography

Abrams v. United States, 250 U.S. 616 (1919).

Abramson, Jeffrey, Christopher Arterton & Gary R. Orren. (1988). *The Electronic Commonwealth: The Impact of New Media Technologies on Democratic Politics*. New York: Basic Books.

Abramson, Jeffrey & Elizabeth Bussiere. (1995). "Free Speech and Free Press: A Communitarian Perspective." In Amitai Etzioni. (Ed.) *New Communitarian Thinking: Persons, Virtues, Institutions, and Communities*. (pp. 218-229). Charlottesville: University Press of Virginia.

Alt, James E. & Kenneth A. Shepsle. (Eds.). (1990). *Perspectives on Positive Political Economy*. New York: Cambridge University Press.

Altschull, J. Herbert. (1995). *Agents of Power: The Media and Public Policy*. 2nd Ed. White Plains, NY: Longman.

Amalgamated Food Employees Union Local 590 v. Logan Valley Plaza, 391 U.S. 308 (1968).

American Civil Liberties Union (ACLU) v. Reno, 929 F.Supp. 824 (E.D.Pa. 1996).

Arterton, F. Christopher. (1987). *Teledemocracy: Can Technology Protect Democracy?* Newbury Park, CA: Sage.

Asai, Sumiko. (1997). "Calculation of Universal Service Costs and Its Significance." *IPTP Monthly Reports*, 106. (available at: http://www.iptp.go.jp/research_e/f_index.html).

Associated Press v. United States, 326 U.S. 1 (1945).

Atkin, David. (1996). "Local and Long Distance Telephony (POTS)." In August E. Grant. (Ed.) *Communication Technology Update*. 5th Ed. (pp. 230-243). Boston: Focal Press.

Aufderheide, Patricia. (1997). "Shifting Policy Paradigms and the Public Interest in the U.S. Telecommunications Act of 1996." *Communication Review*, 2 (2): 259-281.

Babe, Robert E. (1990). *Telecommunications in Canada: Technology, Industry, and Government.* Toronto: University of Toronto Press.

——. (1995). *Communication and the Transformation of Economics: Essays in Information, Public Policy, and Political Economy.* Boulder, CO: Westview Press.

Bagdikian, Ben H. (1985). "The U.S. Media: Supermarket or Assembly Line?" *Journal of Communication*, 35 (3): 97-109.

Baird, Vanessa & Alan Stone. (1998). "Why Privatization? The Case of German Telecommunications." *Social Science Quarterly*, 79 (1): 193-211.

Baker, Edwin. (1989). *Human Liberty and Freedom of Speech.* New York: Oxford University Press.

Bangemann Group. (1994). "Europe and the Global Information Society." (available at: http://www2.echo.lu/eudocs/en/bangemann.html).

Baran, Nicholas. (1996). "Privatization of Telecommunications." *Monthly Review: An Independent Socialist Magazine*, 48 (3): 59-69.

Barber, Benjamin R. (1984). *Strong Democracy: Participatory Politics for a New Age.* Berkeley, CA: University of California Press.

Bates, Stephen. (1996). "The Death of Geography, the Rise of Anonymity, and the Internet." *American Enterprise*, 7 (2): 50-52.

Bauer, Johannes M. (1994). "Conceptual Frameworks for the Design of Telecommunications Policy." In Charles Steinfield, Johannes M. Bauer & Laurence Caby. (Eds.) *Telecommunications in Transition: Policies, Services and Technologies in the European Community.* (pp. 18-35). Thousand Oaks, CA: Sage.

Bauer, Johannes M. & Charles Steinfield. (1994). "Telecommunications Initiatives of the European Communities." In Charles Steinfield, Johannes M. Bauer & Laurence Caby. (Eds.) *Telecommunications in Transition: Policies, Services and Technologies in the European Community.* (pp. 51-70). Thousand Oaks, CA: Sage.

Beer, Lawrence W. (1984). *Freedom of Expression in Japan: A Study in Comparative Law, Politics, and Society.* Tokyo: Kodansha International.

——. (1990). "Freedom of Expression: The Continuing Revolution." *Law and Contemporary Problems*, 53 (2): 39-69.

Berger, Curtis J. (1991). "*Pruneyard* Revisited: Political Activity on Private Lands." *New York University Law Review*, 66 (3): 633-694.

Berman, Jerry & Marc Rotenberg. (1991, January 6). "Free Speech in an Electronic Age." *New York Times*, sec. 3, p. 13.

Berman, Jerry & Daniel J. Weitzner. (1995). "Abundance and User Control: Renewing the Democratic Heart of the First Amendment in the Age of Interactive Media." *Yale Law Journal*, 104 (7): 1619-1637.

——. (1997). "Technology and Democracy." *Social Research*, 64 (3): 1313-1319.

Bhagwat, Ashutosh. (1995). "Of Markets and Media: The First Amendment, the New Mass Media, and the Political Components of Culture." *North Carolina Law Review*, 74 (1): 141-217.

Black, James J. (1997). "Free Speech & The Internet: The Inevitable Move Toward Government Regulation." *Richmond Journal of Law & Technology*, 4 (2). (available at: http//www.Richmond.edu/~jolt/v4i2/black.html).

Blum v. Yaretsky, 457 U.S. 991 (1982).

Blumler, Jay G. (1991). "The New Television Marketplace: Imperatives, Implications, Issues." In James Curran & Michael Gurevitch. (Eds.) *Mass Media and Society.* (pp. 194-215). London: Edward Arnold.

Branscomb, Anne Wells. (1995). "Anonymity, Autonomy, and Accountability: Challenges to the First Amendment in Cyberspace." *Yale Law Journal*, 104 (7): 1639-1679.

Braun, Philipp & Andreas Schaal. (1997). "Federalism, the Nation State and the Global Network: The Case of German Communications Policy." Paper Presented at the Harvard Information Infrastructure Project Conference on the Impact of the Internet on Communications Policy at Harvard University, December 4-5. (available at: http://ksgwww.harvard.edu/iip/iicompol/Papers/Braun-Schaal.html).

Brennan, Geoffrey & James M. Buchanan. (1985). *The Reason of Rules: Constitutional Political Economy.* New York: Cambridge University Press.

British Telecommunications (BT). (1998). *BT World Communications Report 1998/9: The Global Challenge*. (available at: http://www.bt.com/global_reports/1998-99/index.htm).

Brock, Gerald W. (1981). *The Telecommunications Industry: The Dynamics of Market Structure*. Cambridge, MA: Harvard University Press.

Bruce, Robert R., Jeffrey P. Cunard & Mark D. Director. (1986). *From Telecommunications to Electronic Services: A Global Spectrum of Definitions, Boundary Lines, and Structures*. Boston: Butterworths.

Buchanan, James M. (1986). *Liberty, Market and State: Political Economy in the 1980s*. Brighton: Wheatsheaf.

Bundesministerium für Wirtschaft [Federal Ministry of Economics] (BMWi). (1996). "Info 2000: Germany's Way to the Information Society." (available at: http://www.bmwi-info2000.de).

Cable Television Consumer Protection and Competition Act of 1992 (1992 Cable Act). (1992). Public Law No. 102-385, 106 Stat. 1460 (codified at 47 U.S.C.).

Caby, Laurence & Charles Steinfield. (1994). "Trends in the Liberalization of European Telecommunications: Community Harmonization and National Divergence." In Charles Steinfield, Johannes M. Bauer & Laurence Caby. (Eds.) *Telecommunications in Transition: Policies, Services and Technologies in the European Community.* (pp. 36-48). Thousand Oaks, CA: Sage.

Calabrese, Andrew & Mark Borchert. (1996). "Prospects for Electronic Democracy in the United States: Rethinking Communication and Social Policy." *Media, Culture & Society*, 18 (2): 249-268.

Calabrese, Andrew & Donald Jung. (1992). "Broadband Telecommunications in Rural America: Emerging Infrastructures for Residential Service." *Telecommunications Policy*, 16 (3): 225-236.

Carey, James W. (1988). *Communication as Culture: Essays on Media and Society.* New York: Routledge.

Catinat, Michel. (1997). "The Reform of the Telecommunication Regulatory Frameworks: The American and European Approaches." Paper Presented at the Harvard Information Infrastructure Project Conference on the Impact of the Internet on Communications Policy at Harvard University, December 4-5. (available at: http://ksgwww.harvard.edu/iip/iicompol/Papers/Catinat.html).

Chambers, Donald E. (1993). *Social Policy and Social Programs: A Method for the Practical Public Policy Analyst.* 2nd Ed. New York: Macmillan.

Charles, Dan. (1997, May 3). "Net Lobby Praises German Bill." *New Scientist*, 154 (2080): 6.

Civille, Richard. (1995). "The Internet and the Poor." In Brian Kahin & James Keller. (Eds.) *Public Access to the Internet.* (pp. 175-207). Cambridge, MA: MIT Press.

Commission of the European Communities. (1987). "Towards a Dynamic European Economy – Green Paper on the Development of the Common Market for Telecommunications Services and Equipment." COM (87) 290.

Commission of the European Communities. (1988). "Commission Directive of 16 May 1988 on Competition in the Markets in Telecommunications Terminal Equipment." 88/301/EEC; OJ L 131/73. (available at: http://www.ispo.cec.be/infosoc/legreg/docs/ 88301eec.html).

——. (1990). "Commission Directive of 28 June 1990 on Competition in the Markets for Telecommunications Services." 90/388/EEC; OJ L 192/10. (available at: http://www.ispo.cec.be/infosoc/legreg/docs/90388eec.html).

——. (1993). "White Paper on Growth, Competitiveness, and Employment." COM (93) 700. (available at: http://europa.eu.int/en/record/white/ c93700/contents.html).

——. (1994a). "Commission Directive of 13 October 1994 Amending Directive 88/301/EEC and Directive 90/388/EEC in Particular with Regard to Satellite Communications." 94/46/EC; OJ L 268/15. (available at: http://www.ispo.cec.be/infosoc/legreg/docs/9446ec.html).

——. (1994b). "Europe's Way to the Information Society: An Action Plan." COM (94) 347. (available at: http://www2.echo.lu/eudocs/en/com-asc. html).

——. (1996a). "Commission Directive of 16 January 1996 Amending Directive 90/388/EEC with Regard to Mobile and Personal Communications." 96/2/EC; OJ L 20/59. (available at: http://www.ispo.cec.be/infosoc/legreg/docs/962ec.html).

——. (1996b). "Green Paper on the Protection of Minors and Human Dignity in Audiovisual and Information Services." COM (96) 483. (available at: http://www2.echo.lu/legal/en/internet/content/gpen-toc.html).

——. (1996c). "Communication to the European Parliament, the Council, the Economic and Social Committee and the Committee of the Regions: Illegal and Harmful Content on the Internet." COM (96) 487. (available at: http://www2.echo.lu/legal/en/internet/communic.html).

——. (1996d). "Communication to the European Parliament, the Council, the Economic and Social Committee and the Committee of the Regions: Universal Service for Telecommunications in the Perspective of a Fully Liberalised Environment – An Essential Element of the Information Society." COM (96) 73 final. (available at: http://www.ispo.cec.be/ infosoc/telecompolicy/en/d8.htm).

——. (1998). "Communication to the European Parliament, the Council, the Economic and Social Committee and the Committee of the Regions: First Monitoring Report on Universal Service in Telecommunications in the European Union." (available at: http://www.ispo.cec.be/infosoc/ telecompolicy/en/ip98182.html).

Communications Act of 1934 (1934 Act). (1934). 48 Stat. 1064 (codified as amended in scattered sections of 47 U.S.C.).

Compaine, Benjamin M. (1985). "The Expanding Base of Media Competition." *Journal of Communication*, 35 (3): 81-96.

CompuServe Inc. v. Cyber Promotions, Inc., 962 F.Supp. 1015 (S.D.Ohio 1997).

Comstock, George A. (1982). "Information Management and Mass Media: Menace or Myth?" In Jack P. Gibbs. (Ed.) *Social Control: Views from the Social Sciences*. (pp. 205-227). Beverly Hills, CA: Sage.

Cooper, Mark. (1996). "Universal Service: A Historical Perspective and Policies for the Twenty-First Century." (available at: http://www.benton.org/Library/Prospects/prospects.html).

Cornelius v. NAACP Legal Defense & Educational Fund, 473 U.S. 788 (1985).

Council of the European Communities. (1990). "Council Directive of 28 June 1990 on the Establishment of the Internal Market for Telecommunications Services Through the Implementation of Open Network Provision." 90/387/EEC; OJ L 192/1. (available at: http://www.ispo.cec.be/infosoc/legreg/docs/90387eec.html).

——. (1992). "Council Directive of 5 June 1992 on the Application of Open Network Provision to Leased Lines." 92/44/EEC; OJ L 165/27. (available at: http://www.ispo.cec.be/infosoc/legreg/docs/9244eec.html).

——. (1995). "Council Directive of 13 December 1995 on the Application of Open Network Provision to Voice Telephony." 95/62/EC; OJ L 321/6. (available at: http://www.ispo.cec.be/infosoc/legreg/docs/9562ec.html).

——. (1997a). "Common Position Adopted by the Council on 9 June 1997 with a View to Adopting Council Directive 97//EC of the European Parliament and of the Council on the Application of Open Network Provision (ONP) to Voice Telephony and on Universal Service for Telecommunications in a Competitive Environment." 97//EC. (available at: http://www.ispo.cec.be/infosoc/legreg/docs/onpvoice.html).

——. (1997b). "Directive 97/33/EC of the European Parliament and of the Council of 30 June 1997 on Interconnection in Telecommunications with Regard to Ensuring Universal Service and Interoperability Through Application of the Principles of Open Network Provision." OJ L 199/32. (available at: http://www.ispo.cec.be/infosoc/telecompolicy/en/d1-en.htm).

Crandall, Robert W. (1991). *After the Breakup: U.S. Telecommunications in a More Competitive Era*. Washington, DC: Brookings Institution.

Currie, David P. (1994). *The Constitution of the Federal Republic of Germany.* Chicago: University of Chicago Press.

Cyber Promotions, Inc. v. America On-Line, Inc., 948 F.Supp. 436 (E.D.Pa. 1996).

D'Amico, Mary Lisbeth. (1998, May 29). "Germany Troubled by CompuServe Porn Conviction." *Industry Standard*. (available at: http://www.thestandard.net).

D'Amico, Mary Lisbeth & Torsten Busse. (1998, May 28). "Germany Convicts CompuServe Manager in Porn Case." *PC World*. (available at: http://www.pcworld.com).

Dahlman, Carl J. (1979). "The Problem of Externality." *Journal of Law and Economics*, 22 (1): 141-162.

Danielian, Noobar R. (1939). *AT&T: The Story of Industrial Conquest*. New York: Vanguard Press.

Dennis, Everette E. & John V. Pavlik. (1993). "The Coming of Convergence and Its Consequences." In Everette E. Dennis & John V. Pavlik. (Eds.) *Demystifying Media Technology*. (pp. 1-4). Mountain View, CA: Mayfield.

"Der Spiegel and Manager Magazin Release 'Online-Offline,' First Standard Survey of the German Multimedia Market." (1997, February 27). *Two-Ten News Network*. (available at: http://releases.twoten.press.net/releases/date/1997/02/27/SURVEYGerman_Multimedia.html).

Derthick, Martha & Paul J. Quirk. (1985). *The Politics of Deregulation*. Washington, DC: Brookings Institution.

Deutsche Telekom (DT). (1998). *Deutsche Telekom Group Half-Year Report of 1998*. (available at: http://www.dtag.de/dtag/ipl2e/cda/t1).

"Deutsche Telekom Launches Internet Initiative." (1999, January 27). *Deutsche Telekom Latest News*. (available at: http://www.dtag.de/dtag/ipl2e/cda/t1).

Dizard, Wilson P. (1985). *The Coming Information Age: An Overview of Technology, Economics, and Politics*. 2nd Ed. New York: Longman.

Du Boff, Richard B. (1980). "Business Demand and the Development of the Telegraph in the United States, 1844-1860." *Business History Review*, 54 (4): 459-479.

——. (1984). "The Rise of Communications Regulation: The Telegraph Industry, 1844-1880." *Journal of Communication*, 34 (3): 52-66.

Dye, Thomas R. (1998). *Understanding Public Policy*. 9th Ed. Upper Saddle River, NJ: Prentice Hall.

Dyson, Kenneth & Peter Humphreys. (1989). "Deregulating Broadcasting: The Western European Experience." *European Journal of Political Research*, 17 (2): 137-154.

Edelstein, Alex S. (1982). *Comparative Communication Research*. Beverly Hills, CA: Sage.

Esser, Josef & Ronald Noppe. (1996). "Private Muddling Through as a Political Programme? The Role of the European Commission in the Telecommunications Sector in the 1980s." *West European Politics*, 19 (3): 547-562.

"Failures of the German Telecommunications Act of 1996: The Need for Reform." (1997). (available at: http://jhuniverse.hcf.jhu.edu/~aicgsdoc/aicgs/pahefe.htm).

Federal Communications Commission (FCC). (1959). "In the Matter of Allocation of Microwave Frequencies in the Band Above 890 Mc." 27 FCC 359.

——. (1968). "In the Matter of Use of the Carterfone Device in Message Toll Telephone Service." 13 FCC 2d 420, *reconsideration denied*, 14 FCC 2d 571 (1968).

——. (1969). "In Re Applications of Microwave Communications, Inc." 18 FCC 2d 953, *reconsideration denied*, 21 FCC 2d 825 (1970).

——. (1972). "Domestic Communications-Satellite Facilities (DOMSAT)." Second Report and Order, 35 FCC 2d 844, *aff'd sub nom.*

——. (1980). "Computer Inquiry II." 77 FCC 2d 384.

——. (1986). "Computer Inquiry III, Phase I." 104 FCC 2d 958.

Fischer, Frank. (1980). *Politics, Values, and Public Policy: The Problem of Methodology*. Boulder, CO: Westview Press.

Fiss, Owen M. (1986). "Free Speech and Social Structure." *Iowa Law Review*, 71 (5): 1405-1425.

——. (1987). "Why the State?" *Harvard Law Review*, 100 (4): 781-794.

——. (1991). "State Activism and State Censorship." *Yale Law Journal*, 100 (7): 2087-2106.

——. (1996). *The Irony of Free Speech*. Cambridge, MA: Harvard University Press.

Foerstel, Herbert N. (1997). *Free Expression and Censorship in America: An Encyclopedia*. Westport, CT: Greenwood Press.

Fowler, Mark S. & Daniel L. Brenner. (1982). "A Marketplace Approach to Broadcast Regulation." *Texas Law Review*, 60 (2): 207-257.

Fraleigh, Douglas M. & Joseph S. Tuman. (1997). *Freedom of Speech in the Marketplace of Ideas*. New York: St. Martin's Press.

Friedland, Lewis A. (1996). "Electronic Democracy and the New Citizenship." *Media, Culture & Society*, 18 (2): 185-212.

Gandy, Oscar H., Jr. (1993). *The Panoptic Sort: A Political Economy of Personal Information*. Boulder, CO: Westview Press.

Garnham, Nicholas. (1990). *Capitalism and Communication: Global Culture and the Economics of Information*. Newbury Park, CA: Sage.

——. (1992). "Editorial." *Media, Culture & Society*, 14 (3): 339-342.

Gasman, Lawrence. (1998). "Universal Service: The New Telecommunications Entitlements and Taxes." *Cato Policy Analysis*, 310. (available at: http://www.cato.org/pubs/pas/pa-310es.html).

Gershon, Richard A. (1997). *The Transnational Media Corporation: Global Messages and Free Market Competition*. Mahwah, NJ: Lawrence Erlbaum Associates.

Gesetz zur Neuordnung des Postwesens und der Telekommunikation: Postneuordnungsgesetz [Law on Restructuring the Postal Services and Telecommunications] (PTNeuOG). (1994, June 27). BGB1. I S. 2325.

Gesetz zur Neustrukturierung des Post- und Fernmeldewesens und der Deutschen Bundespost [Law on Structuring the Postal and Telecommunications Services and the Federal Postal System] (PostStruktG). (1989, June 8). BGB1. I S.1026. (F.R.G.).

Giese, Mark. (1996). "From ARPA to the Internet: A Cultural Clash and Its Implications in Framing the Debate on the Information Superhighway." In Lance Strate, Ronald Jacobson & Stephanie B. Gibson. (Eds.) *Communication and Cyberspace: Social Interaction in an Electronic Environment*. (pp. 123-141). Cresskill, NJ: Hampton Press.

Girard, Kim. (1997a). "Telecom Reform Fails to Pay off for Users." *Computerworld*, 31 (6): 4.

——. (1997b). "Freewheeling Days to End for Net Providers." *Computerworld*, 31 (20): 2.

Golding, Peter. (1996). "World Wide Wedge: Division and Contradiction in the Global Information Infrastructure." *Monthly Review: An Independent Socialist Magazine*, 48 (3): 70-85.

Golding, Peter & Graham Murdock. (1991). "Culture, Communications, and Political Economy." In James Curran & Michael Gurevitch. (Eds.) *Mass Media and Society*. (pp. 15-32). London: Edward Arnold.

Gore, Al, Jr. (1993, December 21). "Remarks by the Vice President at the National Press Club." Washington, DC.

Grossman, Lawrence K. (1995). *The Electronic Republic: Reshaping Democracy in the Information Age*. New York: Penguin Books.

Grundgesetz [Basic Law] (F.R.G.). (available at: http://www.jura.uni-sb.de/law/GG/gg0.html).

Hacker, Kenneth L. (1996). "Missing Links in the Evolution of Electronic Democratization." *Media, Culture & Society*, 18 (2): 213-232.

Hadden, Susan G. & Edward Lenert. (1995). "Telecommunications Networks are Not VCRs: The Public Nature of New Information Technologies for Universal Service." *Media, Culture & Society*, 17 (1): 121-140.

Hague v. Committee for Industrial Organization, 307 U.S. 496 (1939).

Hamelink, Cees J. (1994). *The Politics of World Communication: A Human Rights Perspective*. Thousand Oaks, CA: Sage.

Hancher, Leigh & Michael Moran. (1989). "Introduction: Regulation and Deregulation." *European Journal of Political Research*, 17 (2): 129-136.

Hardach, Karl. (1977). "Germany 1914-1970." In Carlo M. Cipolla. (Ed.) *The Fontana Economic History of Europe Vol. 6: Contemporary Economies – 1*. (pp. 180-265). New York: Barnes & Noble.

Harvey, Lee. (1990). *Critical Social Research*. London: Unwin Hyman.

Heffron v. International Society for Krishna Consciousness, Inc., 452 U.S. 640 (1981).

Herman, Edward S. & Robert W. McChesney. (1997). *The Global Media: The New Missionaries of Corporate Capitalism*. London: Cassell.

Hills, Jill. (1986). *Deregulating Telecoms: Competition and Control in the United States, Japan, and Britain*. London: Frances Pinter.

——. (1991). *The Democracy Gap: The Politics of Information and Communication Technologies in the United States and Europe*. New York: Greenwood.

Hoffmann-Riem, Wolfgang. (1984). "Policy Research on Telecommunications in West Germany." In Vincent Mosco. (Ed.) *Policy Research in Telecommunication*. (pp. 55-70). Norwood, NJ: Ablex.

Horwitz, Robert B. (1989). *The Irony of Regulatory Reform: The Deregulation of American Telecommunications*. New York: Oxford University Press.

——. (1991). "The First Amendment Meets Some New Technologies: Broadcasting, Common Carriers, and Free Speech in the 1990s." *Theory and Society*, 20 (1): 21-72.

Hudgens v. NLRB, 424 U.S. 507 (1976).

Hudis, Mark. (1994). "Interactive: Pioneers are Making Inroads into Interactive Territory." *MediaWeek*, 4 (36): 40-41.

Hudson, Heather E. (1997). *Global Connections: International Telecommunications Infrastructure and Policy*. New York: Van Nostrand Reinhold.

Hush-A-Phone Corp. v. United States, 238 F.2d 266 (D.C. Cir. 1956).

Information Infrastructure Task Force (IITF). (1993). *The National Information Infrastructure: Agenda for Action*. Washington, DC: United States Department of Commerce. (available at: http://sunsite.unc.edu/nii/NII-Agenda-for-Action.html).

Informations- und Kommunikationsdienste-Gesetz [Information and Communication Services Act] (IuKDG). (1997, June 13). (available at: http://www.iid.de/iukdg/iukdge.html).

"In the Shark Pond." (1998, January 3). *The Economist*, 345 (8049): 59-60.

Ito, Youichi. (1986). "Telecommunications and Industrial Policies in Japan: Recent Developments." In Marcellus S. Snow. (Ed.) *Marketplace for Telecommunications: Regulation and Deregulation in Industrialized Democracies*. (pp. 201-230). New York: Longman.

Jansen, Sue Curry. (1988). *Censorship: The Knot That Binds Power and Knowledge*. New York: Oxford University Press.

"Japanese Internet Commerce to be Worth $126 Billion." (1999). *Japan Internet Report*, 35. (available at: http://www.tkai.com).

Jessop, Bob. (1990). *State Theory: Putting the Capitalist State in Its Place.* University Park: Pennsylvania State University Press.

Johnston, Margret. (1997). "Deutsche Telekom Signs Interconnect Agreement with Isis." *InfoWorld*, 19 (37): 64.

Jussawalla, Meheroo. (1995). "Telecommunications Privatization and Capital Formation in the ASEAN." In Bella Mody, Johannes M. Bauer & Joseph D. Straubhaar. (Eds.) *Telecommunications Politics: Ownership and Control of the Information Highway in Developing Countries*. (pp. 165-178). Mahwah, NJ: Lawrence Erlbaum Associates.

Kahin, Brian. (1995). "The Internet and the National Information Infrastructure." In Brian Kahin & James Keller. (Eds.) *Public Access to the Internet*. (pp. 3-23). Cambridge, MA: MIT Press.

——. (1997). "The U.S. National Information Infrastructure: The Market, the Net, and the Virtual Project." In Brian Kahin & Ernest J. Wilson III. (Eds.) *National Information Infrastructure Initiatives: Vision and Policy Design*. (pp. 150-189). Cambridge, MA: MIT Press.

Kahn, Alfred E. (1983). "The Passing of the Public Utility Concept: A Reprise." In Eli M. Noam. (Ed.) *Telecommunications Regulation Today and Tomorrow*. (pp. 3-37). New York: Law & Business.

Kalba, Kas. (1988). "Opening Japan's Telecommunications Market." *Journal of Communication*, 38 (1): 96-106.

Katz, Michael L. (1997). "Ongoing Reform of U.S. Telecommunications Policy." *European Economic Review*, 41 (3-5): 681-690.

Keane, John. (1991). *The Media and Democracy.* Cambridge, MA: Polity Press.

Kennedy, Melanie A. (1996). "Information Superhighway: Parental Regulation - The Best Alternative." *University of Louisville Journal of Family Law,* 35 (3): 575-593.

"Keyboard Allergy Dangerous to Financial Health." (1999). *Japan Internet Report,* 34. (available at: http://www.tkai.com).

Kim, Jae-Young. (1998a). "Universal Service and Internet Commercialization: Chasing Two Rabbits at the Same Time." *Telecommunications Policy,* 22 (4/5): 281-288.

——. (1998b). "Filtering Software vs. Free Speech: Towards a Right Option for Regulation Internet Pornography." Paper Presented in the Communication Technology Division at the Annual Convention of the Broadcast Education Association, Las Vegas, April 3-6.

——. (1999). "First Amendment Rights on the Internet: A Snowball's Chance in Hell?" In Lester J. Pourciau. (Ed.) *Ethics and Electronic Information in the Twenty-First Century.* (pp. 76-92). West Lafayette, IN: Purdue University Press.

"Kingsbury Commitment." (1913, December 19). Letter from N.C. Kingsbury, AT&T to J.C. McReynolds, Attorney General, Department of Justice.

Kokusai Denshin Denwa Company Ltd. Law. (1953). Law No. 301 of August 7. (available at: http://www.mpt.go.jp/policyreports/english/laws/Kdd_law.html).

Kommers, Donald P. (1997). *The Constitutional Jurisprudence of the Federal Republic of Germany.* 2nd Ed. Durham: Duke University Press.

Koshu Denki Tsushinho Law [Public Telecommunications Law]. (1953). Law No. 97.

Krantz, Michael. (1997, August 11). "Censor's Sensibility." *Time.* 48-49.

Krattenmaker, Thomas G. (1996). "The Telecommunications Act of 1996." *Federal Communications Law Journal,* 49 (1): 1-49.

Kress, Carl B. (1997). "The 1996 *Telekommunikationsgesetz* and the Telecommunications Act of 1996: Toward More Competitive Markets in Telecommunications in Germany and the United States." *Federal Communications Law Journal,* 49 (3): 551-619.

Krynicky v. University of Pittsburgh, 742 F.2d 94 (1984).

Kubicek, Herbert. (1997). "Multimedia: Germany's Third Attempt to Move to an Information Society." In Brian Kahin & Ernest J. Wilson III. (Eds.) *National Information Infrastructure Initiatives: Vision and Policy Design.* (pp. 387-423). Cambridge, MA: MIT Press.

"Launching Deutsche Telekom." (1996, October 26). *Economist*, 341 (7989): 73-74.

Leopold, George. (1996, June 10). "Japan Lags on Internet." *Electronic Engineering Times*, 905: 18-19.

Lloyd Corp. v. Tanner, 407 U.S. 551 (1972).

Lowenstein, Ralph L. & John C. Merrill. (1990). *Macromedia: Mission, Message, and Morality*. New York: Longman.

McChesney, Fred S. (1987). "Rent Extraction and Rent Creation in the Economic Theory of Regulation." *Journal of Legal Studies*, 16 (1): 101-118.

McChesney, Robert W. (1993). *Telecommunications, Mass Media, and Democracy: The Battle for the Control of U.S. Broadcasting, 1928-1935*. New York: Oxford University Press.

——. (1996a). "The Global Struggle for Democratic Communication." *Monthly Review: An Independent Socialist Magazine*, 48 (3): 1-20.

——. (1996b). "The Internet and U.S. Communication Policy-Making in Historical and Critical Perspective." *Journal of Communication*, 46 (1): 98-124.

——. (1997). "Digital Highway Robbery: Where is the 'Competition' the Telecommunications Act was Supposed to Provide?" *Nation*, 264 (15): 22-24.

McChesney, Robert W., Ellen Meiksins Wood & John Bellamy Foster. (Eds.). (1998). *Capitalism and the Information Age: The Political Economy of the Global Communication Revolution*. New York: Monthly Review Press.

McCormick, Robert E. (1989). "A Review of the Economics of Regulation: The Political Process." In Roger E. Meiners & Bruce Yandle. (Eds.) *Regulation and the Reagan Era: Politics, Bureaucracy and the Public Interest*. (pp. 16-37). New York: Holmes & Meier.

McKeesport Hospital v. Accreditation Council for Graduate Medical Ed., 24 F.3d 519 (3rd Cir. 1994).

McQuail, Denis. (1994). *Mass Communication Theory: An Introduction*. 3rd Ed. Thousand Oaks, CA: Sage.

Mahon, Rianne. (1980). "Regulatory Agencies: Captive Agents or Hegemonic Apparatus." In J. Paul Grayson. (Ed.) *Class, State, Ideology, and Change: Marxist Perspectives on Canada*. (pp. 154-168). Toronto: Holt, Rinehart, and Winston.

Mansell, Robin. (1993). *The New Telecommunications: A Political Economy of Network Evolution*. Thousand Oaks, CA: Sage.

Mark v. Borough of Hatboro, 51 F.3d 1137 (3rd Cir. 1995).

Marsh v. Alabama, 326 U.S. 501 (1946).

Martinez, Carlos E. (1992). "Early Lessons of Latin American Privatizations." *Suffolk Transnational Law Journal*, 15 (2): 468-513.

Marushin Industries, Inc. v. Japan, Judgment of November 22, 1972, Supreme Court, Grand Bench, 26 Keishu 586.

Matsui, Shigenori. (1994). "*Lochner v. New York* in Japan: Protecting Economic Liberties in a Country Governed by Bureaucrats." In Philip S. C. Lewis. (Ed.) *Law and Technology in the Pacific Community.* (pp. 199-261). Boulder, CO: Westview Press.

Mediendienste-Staatsvertrag [Media Services Treaty] (MdStV). (1997, April 14). (available at: http://www.uni-muenster.de/Jura.itm/netlaw/mdsvnw.html).

Meiklejohn, Alexander. (1960). *Political Freedom: The Constitutional Powers of the People.* New York: Harper & Brothers.

Meyer, Carlin. (1995). "Reclaiming Sex from the Pornographers: Cybersexual Possibilities." *Georgetown Law Journal*, 83 (5): 1969-2008.

Meyerson, Michael I. (1997). "Ideas of the Marketplace: A Guide to the 1996 Telecommunications Act." *Federal Communications Law Journal*, 49 (2): 251-288.

"Microsoft Firing Raises Alarm over Online Free Speech." (1993, July 15). *The Plain Dealer*, 1.

Miller v. California, 413 U.S. 15 (1973).

Miller, Edythe S. (1996). "Economic Regulation and New Technology in the Telecommunications Industry." *Journal of Economic Issues*, 30 (3): 719-735.

Miller, Steven E. (1996). *Civilizing Cyberspace: Policy, Power, and the Information Superhighway.* New York: ACM Press.

Ministry of International Trade and Industry Committee. (1997). "Interim Report on Technological, Legal and Institutional Issues on Electronic Commerce." (available at: http://www.miti.go.jp/info-e/c-menu-e.html).

Ministry of Posts and Telecommunications (MPT). (1996). "Report Submitted by Study Group for Research into Universal Services and Rates in the Multimedia Age." (available at: http://www.mpt.go.jp/pressrelease/english/telecomm/news7-7-2.html).

——. (1997). "Internet Open Salon '*Jiyu-Kattatsu*' Established." (available at: http://www.mpt.go.jp/pressrelease/english/general/news7-20-4.html).

——. (1998). *White Paper: Communications in Japan, 1998.* (available at: http://www.mpt.go.jp/policyreports/english/papers/WhitePaper1998.pdf).

Ministry of Posts and Telecommunications Telecommunications Bureau. (1996). "Flow of Information on the Internet: Report of the Study Group for the Advancement of the Condition for the Use of Telecommunications." (available at: http://www.mpt.go.jp/policyreports/english/group/Internet/contents.html).

——. (1997). "The Rules for the Flow of Information on the Internet: Report of the Study Group on the Rules for the Flow of Information in the Telecommunications Services." (available at: http://www.mpt.go.jp/policyreports/english/telecommunications/rules_reportidx_e.html).

Ministry of Posts and Telecommunications Telecommunications Council. (1994). "Reforms Toward the Intellectually Creative Society of the 21st Century: Program for the Establishment of High-Performance Info-Communications Infrastructure." (available at: http://www.mpt.go.jp/policyreports/english/telecouncil/Report1993No5/contents.html).

——. (1995). "For Achieving Globalization of an 'Intellectually Creative Society': Interim Report." (available at: http://www.mpt.go.jp/policyreports/english/telecouncil/Interim_Report/index.html).

——. (1996). "Status of Nippon Telegraph and Telephone Corporation." (available at: http://www.mpt.go.jp/policyreports/english/telecouncil/NTT/Council-NTTs.html).

——. (1997). "Vision 21 for Info-Communications: Policies to be Promoted as We Approach the 21st Century and the Specific Socioeconomic Conditions That Will Emerge." (available at: http://www.mpt.go.jp/policyreports/english/telecouncil/v21-9706/v21-9706-e.html).

Mody, Bella & Lai-Si Tsui. (1995). "The Changing Role of the State." In Bella Mody, Johannes M. Bauer & Joseph D. Straubhaar. (Eds.) *Telecommunications Politics: Ownership and Control of the Information Highway in Developing Countries*. (pp. 179-198). Mahwah, NJ: Lawrence Erlbaum Associates.

Mosco, Vincent. (1982). *Pushbutton Fantasies: Critical Perspectives on Videotex and Information Technology*. Norwood, NJ: Ablex.

——. (1984). "Introduction." In Vincent Mosco. (Ed.) *Policy Research in Telecommunications*. (pp. ix-xiii). Norwood, NJ: Ablex.

——. (1988). "Toward a Theory of the State and Telecommunications Policy." *Journal of Communication*, 38 (1): 107-124.

——. (1989). *The Pay-per Society: Computers and Communication in the Information Age – Essays in Critical Theory and Public Policy*. Norwood, NJ: Ablex.

Mosco, Vincent. (1996). *The Political Economy of Communication: Rethinking and Renewal*. Thousand Oaks, CA: Sage.

Mosco, Vincent & Janet Wasko. (Eds.). (1988). *The Political Economy of Information*. Madison: University of Wisconsin Press.

Mueller, Milton. (1995). "Why Communications Policy is Passing 'Mass Communication' by: Political Economy as the Missing Link." *Critical Studies in Mass Communication*, 12 (4): 457-472.

——. (1997a). "Universal Service and the Telecommunications Act: Myth Made Law." *Communications of the ACM*, 40 (3): 39-47.

——. (1997b). "Telecommunications Access in the Age of Electronic Commerce: Toward a Third-Generation Universal Service Policy." *Federal Communications Law Journal*, 49 (3): 655-673.

Naoe, Shigehiko. (1994). "Japan's Telecommunications Industry: Competition and Regulatory Reform." *Telecommunications Policy*, 18 (8): 651-657.

National Telecommunications and Information Administration (NTIA). (1999). "Falling Through the Net: Defining the Digital Divide." (available at: http://www.ntia.doc.gov/ntiahome/fttn99/contents.html).

Naughton, Edward J. (1992). "Is Cyberspace a Public Forum? Computer Bulletin Boards, Free Speech, and State Action." *Georgetown Law Journal*, 81 (2): 409-441.

Neumann, Holger. (1998). "First Cases Under the New German Telecommunications Act: Competitor's Access to Telekom's Customer Access Lines." (available at: http://www.bws.de/library/0015/0015.html).

New Jersey Coalition Against War in the Middle East v. J.M.B. Realty Corp., 650 A.2d 757 (N.J. 1994).

New York Times, Co. v. Sullivan, 376 U.S. 254 (1964).

Nippon Denshin Denwa Kabushiki Kaisha Law. (1984). Law No. 85 of December 25. (available at: http://www.mpt.go.jp/policyreports/english/laws/Ntt_law.html).

Noam, Eli M. (1992). *Telecommunications in Europe*. New York: Oxford University Press.

Noam, Eli M. & Richard A. Kramer. (1994). "Telecommunications Strategies in the Developing World: A Hundred Flowers Blooming or Old Wine in New Bottles?" In Charles Steinfield, Johannes M. Bauer & Laurence Caby. (Eds.) *Telecommunications in Transition: Policies, Services and Technologies in the European Community*. (pp. 272-286). Thousand Oaks, CA: Sage.

Nohe, Richard E. (1996). "A Different Time, a Different Place: Breaking up Telephone Companies in the United States and Japan." *Federal Communications Law Journal*, 48 (2): 307-340.

Nora, Simon & Alain Minc. (1980). *The Computerization of Society: A Report to the President of France*. Cambridge, MA: MIT Press.

Oniki, Hajime. (1993). "Impacts of the 1985 Reform of Japan's Telecommunications Industry on NTT." In Meheroo Jussawalla. (Ed.) *Global Telecommunications Policies: The Challenge of Change*. (pp. 69-92). Westport, CT: Greenwood Press.

Organization for Economic Cooperation and Development (OECD). (1999). *Communications Outlook 1999*. (available at: http://www.oecd.org).

Pavlik, John V. (1996). *New Media Technology: Cultural and Commercial Perspectives*. Boston: Allyn & Bacon.

Perritt, Henry H., Jr. (1995). "Access to the National Information Infrastructure." *Wake Forrest Law Review*, 30 (1): 51-103.

Perry Education Association v. Perry Local Educators' Association, 460 U.S. 37 (1983).

Pigou, Arthur C. (1920). *The Economics of Welfare*. London: Macmillan.

Pool, Ithiel de Sola. (1983). *Technologies of Freedom*. Cambridge, MA: Belknap Press.

Pridgen, Dee. (1997). "How Will Consumers be Protected on the Information Superhighway?" *Land & Water Law Review*, 32 (1): 237-255.

"Prime Minister's Policy Speech." (1996). *Japan Echo*, 23 (1): 62-65.

"Protecting Against Porn." (1997, October 2). *Orange County Register*, B10.

Pruneyard Shopping Center v. Robins, 447 U.S. 74 (1980).

Rainey, R. Randall & William Rehg. (1996). "The Marketplace of Ideas, the Public Interest, and Federal Regulation of the Electronic Media: Implications of Habermas' Theory of Democracy." *Southern California Law Review*, 69 (6): 1923-1987.

Read, William H. & Jan L. Youtie. (1996). *Telecommunications Strategy for Economic Development*. Westport, CT: Praeger.

Red Lion Broadcasting Co. v. Federal Communications Commission, 395 U.S. 367 (1969).

Reno v. American Civil Liberties Union (ACLU), 117 S.Ct. 2329 (1997).

Reuben-Cooke, Wilhelmina M. (1993). "Rethinking Legal and Policy Paradigms." In Charles M. Firestone. (Ed.) *Television for the 21st Century: The Next Wave*. (pp. 55-77). Queenstown, MD: Aspen Institute.

Rheingold, Howard. (1994). *The Virtual Community: Homesteading on the Electronic Frontier*. New York: HarperPerennial.

Rogers, Everett M. (1986). *Communication Technology: The New Media in Society*. New York: Free Press.

Rosen, Dan. (1997). "Surfing the *Sento*." *Berkeley Technology Law Journal*, 12 (1): 213-230.

Ryan, Daniel J. (1997). "The Evolving Telecommunications Environment in Japan." In Daniel J. Ryan. (Ed.) *Privatization and Competition in Telecommunications: International Developments*. (pp. 21-32). Westport, CT: Greenwood Press.

Sawhney, Harmeet. (1994). "Universal Service: Prosaic Motives and Great Ideals." *Journal of Broadcasting & Electronic Media*, 38 (4): 375-395.

——. (1996). "Information Superhighway: Metaphors as Midwives." *Media, Culture & Society*, 18 (2): 291-314.

Sayer, Peter & Sarah Deveaux. (2000, July 28). "Jurisdiction in Cyberspace: For Internet Users, the Battle Between Freedom and Order Rages and the Stakes Get Higher." *PC World*. (available at: http://www.pcworld.com).

Schement, Jorge Reina, Rebecca R. Pressman & Laurance Povich. (1995). "Transcending Access Toward a New Universal Service." Paper Presented at the Conference on Universal Service in Context: A Multidisciplinary Perspective at New York Law School, December 6. (available at: http://www.benton.org/Policy/Uniserv/Conference/transcend.html).

Schiller, Dan. (1982). *Telematics and Government*. Norwood, NJ: Ablex.

——. (1994). "From Culture to Information and Back Again: Commoditization as a Route to Knowledge." *Critical Studies in Mass Communication*, 11 (1): 93-115.

Schiller, Herbert I. (1994). "Electronic Highway to Where?" *National Forum: The Phi Kappa Phi Journal*, 74 (2): 19-21.

——. (1996). *Information Inequality: The Deepening Social Crisis in America*. New York: Routledge.

Shaw, James. (1998). *Telecommunications Deregulation*. Boston: Artech House.

Shenk, Joshua Wolf. (1995). "The Robber Barons of the Information Highway." *Washington Monthly*, 27 (6): 17-22.

Sheridan, Kate. (1998, December 19). "Censorship: AOL Shuts Down 23 Irish Forum Sites, Strands Thousands of Irish Customers." *The Irish People*. (available at: http://www.inac.org/irishpeople/index.html).

Smith, Stephen A. (1994). "Communication and the Constitution in Cyberspace." *Communication Education*, 43 (2): 87-101.

Smythe, Dallas W. (1957). *The Structure and Policy of Electronic Communication*. Urbana: University of Illinois Press.

Snow, Marcellus S. (1995). "The AT&T Divestiture: A 10-Year Retrospective." In Donald M. Lamberton. (Ed.) *Beyond Competition: The Future of Telecommunications*. (pp. 207-226). Amsterdam: Elsevier.

Solomon, Deborah. (1997, June 26). "Phone Firms Challenge Federal Communications Commission's Internet Order." *Detroit Free Press.*

Steinglass, David E. (1996). "Extending *Pruneyard*: Citizens' Right to Demand Public Access Cable Channels." *New York University Law Review*, 71 (4): 1113-1160.

Stigler, George J. (1971). "The Theory of Economic Regulation." *Bell Journal of Economics and Management Science*, 2 (1): 3-21.

Sugaya, Minoru. (1997). "Advanced Universal Service in Japan." *Telecommunications Policy*, 21 (2): 177-184.

Sunstein, Cass R. (1988). "Beyond the Republican Revival." *Yale Law Journal*, 97 (8): 1539-1590.

——. (1992). "Free Speech Now." *University of Chicago Law Review*, 59 (1): 255-316.

——. (1995). "The First Amendment in Cyberspace." *Yale Law Journal*, 104 (7): 1757-1804.

Sussman, Gerald. (1997). *Communication, Technology, and Politics in the Information Age*. Thousand Oaks, CA: Sage.

Syracuse Peace Council v. Federal Communications Commission, 867 F. 2d 654 (D.C. Cir. 1989).

Syracuse Peace Council v. WTVH, 2 F.C.C.R. 5043 (1987).

Taviss, Michael L. (1992). "Dueling Forums: The Public Forum Doctrine's Failure to Protect the Electronic Forum." *University of Cincinnati Law Review*, 60 (3): 757-795.

Tehranian, Majid. (1990). *Technologies of Power: Information Machines and Democratic Prospects*. Norwood, NJ: Ablex.

Telecommunications Act of 1996 (1996 Act). (1996). Public Law No. 104-104, 110 Stat. 56 (codified in scattered sections of 47 U.S.C.). (available at: http://www.fcc.gov/Reports/tcom1996.txt).

Telecommunications Business Law. (1984). Law No. 86 of December 25. (available at: http://www.mpt.go.jp/policyreports/english/laws/Tb_index.html).

Telekommunikationsgesetz [Telecommunications Law] (TKG). (1996, August 1). BGB1. I S. 1120. (F.R.G.). (available at: http://www.regtp.de/en/index.html).

Temin, Peter. (1987). *The Fall of the Bell System: A Study in Prices and Politics*. New York: Cambridge University Press.

"T-Online Cuts Rates by More Than 50 Percent." (1999, January 27). *Deutsche Telekom Latest News*. (available at: http://www.dtag.de/dtag/ipl2e/cda/t1).

Trebing, Harry M. (1984). "Successes and Failures of Policy-Oriented Research in the Public Utility Industries." In Vincent Mosco. (Ed.) *Policy Research in Telecommunications*. (pp. 3-17). Norwood, NJ: Ablex.

Tunstall, Jeremy. (1986). *Communications Deregulation: The Unleashing of America's Communications Industry.* New York: Basil Blackwell.

Turner Broadcasting Systems, Inc. v. FCC, 117 S.Ct. 1174 (1997).

Twenhafel, David, John Horrigan, Amy M. Korzick & David McCarty. (1989). "Introduction." In Jurgen Schmandt, Frederick Williams & Robert H. Wilson. (Eds.) *Telecommunications Policy and Economic Development.* (pp. 1-16). New York: Praeger.

United States v. AT&T, 552 F.Supp. 131 (D.D.C. 1982), *aff'd sub nom.*, Maryland v. United States, 460 U.S. 1001 (1983).

Ure, John & Araya Vivorakij. (1997). "Privatization of Telecoms in Asia." In Daniel J. Ryan. (Ed.) *Privatization and Competition in Telecommunications: International Developments*. (pp. 1-20). Westport, CT: Greenwood Press.

Vaillancourt, Pauline Marie. (1986). *When Marxists Do Research.* Westport, CT: Greenwood Press.

Waverman, Leonard & Esen Sirel. (1997). "European Telecommunications Markets on the Verge of Full Liberalization." *Journal of Economic Perspectives*, 11 (4): 113-126.

Weil, Nancy. (1998, May 20). "Telecom Merger Trends Not Good for Users, Study Finds." *IDG News Service*. (available at: http://www.idgnet.com/english/crd_users_20474.html).

Wenders, John T. (1989). "Deregulating Telecommunications." In Roger E. Meiners & Bruce Yandle. (Eds.) *Regulation and the Reagan Era: Politics, Bureaucracy and the Public Interest.* (pp. 104-131). New York: Holmes & Meier.

West, Livia Solange. (1996). "Deregulating Telecommunications: The Conflict Between Competition and Universal Service." *DePaul Business Law Journal*, 9 (1): 159-188.

West, Joel, Jason Dedrick & Kenneth L. Kraemer. (1997). "Back to the Future: Japan's NII Plans." In Brian Kahin & Ernest J. Wilson III. (Eds.) *National Information Infrastructure Initiatives: Vision and Policy Design*. (pp. 61-111). Cambridge, MA: MIT Press.

"Will Commercial Forces Overwhelm Needs of Public-Interest Journalism?" (1994). *Nieman Reports*, 48 (2): 43-50.

Willis-Graham Act. (1921). Public Law No. 15, 42 Stat. 27.

Wilson, Kevin G. (1992). "Deregulating Telecommunications and the Problem of Natural Monopoly: A Critique of Economics in Telecommunications Policy." *Media, Culture & Society*, 14 (3): 343-368.

Winseck, Dwayne Roy. (1993). "A Study of the (De)Regulatory Process in Canadian Telecommunications: Labour Struggles and the Public Interest." Unpublished Ph.D. Diss. University of Oregon.

Wykoff, Frank C. (1997). "Privatization – The Next American Revolution." *Nieman Reports*, 51 (2): 14-15.

Zelezny, John D. (1997). *Communications Law: Liberties, Restraints, and the Modern Media*. 2nd Ed. Belmont, CA: Wadsworth Publishing.

Index